Genomics of Cultivated Palms

Genomics of Cultivated Palms

V. Arunachalam
Horticulture Section
ICAR Research Complex for Goa
(Indian Council of Agricultural Research)
Old Goa, Goa, India

ELSEVIER

AMSTERDAM • BOSTON • HEIDELBERG • LONDON • NEW YORK • OXFORD
PARIS • SAN DIEGO • SAN FRANCISCO • SINGAPORE • SYDNEY • TOKYO

Elsevier
32 Jamestown Road, London NW1 7BY
225 Wyman Street, Waltham, MA 02451, USA

First edition 2012

British Library Cataloguing-in-Publication Data
A catalog record for this book is available from the British Library

Library of Congress Cataloging-in-Publication Data
A catalog record for this book is available from the Library of Congress

ISBN: 978-0-12-387736-9

For information on all Elsevier publications
visit our website at elsevierdirect.com

This book has been manufactured using Print on Demand technology. Each copy is produced to order and is limited to black ink. The online version of this book will show color figures where appropriate.

Dedicated to my parents V.S. Vadivel and Roopavathi Vadivel

Contents

Preface

There is a big gap in the knowledge of molecular biology resources available in palms. I wrote the book on genomics of cultivated palms mainly to fill this gap. Palms are major cultivated crops but have received little scientific attention except for oil palm. Recently, the whole genome sequence for date palm became available in the public domain. Plenty of ESTs, linkage mapping, and markers have been developed in coconut, oil palm, and date palm.

The book will serve as a reference text to students, teachers, and researchers working with palms. I reviewed all the literature available so far on genomics for palms and have compiled it here. It will be of immense use to a researcher who intends to perform a genomics project on palms, as it provides a handy review of the literature, including gaps in the research, primers that work in specific palms, and so on. I have covered some new topics such as the pharmacogenomics of areca and the structural biology of hirsutellin, which have not been reported on in other books covering palm biotechnologies.

The Internet, library, and papers published on palms supplied the details for my study. I wrote the book with the compiled information as tables on samples, using materials and methods employed and conclusion drawn.

I found very little research on most palms except some work on molecular phylogeny. I might have overlooked some important papers due to my shortsightedness and other limitations in accessing full text papers. Currently, much work is being initiated in palms, which will be covered in forthcoming editions.

I am wholly responsible for the contents of the book, and not the organizations for whom I have worked and with whom I am currently working. I hope the readers find it a useful text for research.

Acknowledgments

It is a delight to thank the people who helped me to produce this book. I thank my teachers at Tamil Nadu Agricultural University Coimbatore, Tamil Nadu, India, and the Indian Agricultural Research Institute, Pusa, New Delhi, India, for providing basic knowledge and shaping my professional life.

I am grateful to my colleagues at Central Plantation Crops Research Institute, Kasaragod, Kerala, India, who provided a competitive atmosphere for learning the genes and genome of cultivated palms.

I am grateful to the CERA (Consortium of Educational Resources in Agriculture) Initiative of the National Agricultural Innovation Project of the ICAR (Indian Council of Agricultural Research) for providing access to full-text journal articles. I would like to express my sincere gratitude to the staff and scientists of the Horticulture section, ICAR Research Complex for Goa, especially Dr. M. Thangam and Mr. Rahul Kulkarni for taking photographs. My thanks is due to Dr. P. Murugesan, Directorate of Oil Palm Research, for the plates on cross-sectional views of oil palm fruits.

My thanks are due to Mr. Paulprasad Chandramohan, Mr. Greg Harris, Ms. Lisa Tickner, Ms. Erin Hill-Parks, Ms. Tracey Miller, and Sujatha Thirugnana Sambandam of the Elsevier insights team for the invitation to publish and for regular e-mails to help me complete the task. I also thank Sivakami Ramanathan of MPS Limited, Chennai, India, for her support in editing the volume. I am grateful to my parents (Vadivel and Roopavathi), who gave me a basic education and moral support. My thanks also goes to my family members—my wife, Devasena, and children, Elavarasi and Aksaya—for supporting me and providing a loving environment in my life.

Finally, my utmost thanks is to God, for giving me the courage to face the complexities of life and complete this project successfully.

Abbreviations

8-OH-dG	8-Hydroxydeoxyguanosine
ABA	Abscisic acid
AFLP	Amplified fragment length polymorphism
AGA1	Agamous-1
avβ6	alphavbeta6 integrin
AYD	Andaman Yellow Dwarf
Bg	*Bactris gasipaes*
bp	base pair
CAD	Cinnamyl alcohol-NADPH-dehydrogenase
CDK	Cyclic dependent kinase
cDNA	complimentary DNA
CIRAD	Centre de coopération internationale en recherche agronomique pour le dévelopement
cM	centiMorgan
Cp	Chloroplast
CPCRI RS	Central Plantation Crops Research Institute Regional Station
CRD	Cameroon Red Dwarf
Db	Database
DNA	Deoxyribo nucleic acid
EcoRI	*Escherichia coli* Restriction endonuclease I
Eg	*Elaeis guineensis*
Eo	*Elaeis oleifera*
EST	Expressed sequence tag
F1	Filial generation 1
GDR	Genome database for Rosaceae
HtA	Hirstullein A
ISSR	Inter simple sequence repeat
ISTR	Inter sequence tagged repeat
ITS	Internal transcribed region
kb	kilobase
KGF-1	Keratinocyte growth factor
KNOX	Knotted homeobox gene
KYD	Kulasekaram Yellow Dwarf
LC MS	Liquid chromatography–mass spectrometry
LIS	Legume Information System
LYD	Lethal yellowing disease
MADS	MCM1, agamous, deficiens, SRF
MAPK	Mitogen-activated protein kinase
Mb	Megabase
MS	Malate synthase
MseI	Micrococcus species restriction endonuclease I

Mt3	Metallothionein 3
MYD	Malayan Yellow Dwarf
NAC	No apical meristem
NMR	Nuclear magnetic resonance
OSCC	Oral squamous cell carcinoma
OSF	Oral submucous fibrosis
PAL	Phenylalanine ammonia lyase
PHYB	Phytochrome B
POD	Peroxidase
PPO	Polyphenol oxidase
PRK	Phospho-ribulose kinase
QTL	Quantitative trait loci
rANE	Ripe arecanut extract
RAPD	Random amplified polymorphic DNA
RFLP	Restricted fragment length polymorphism
RGA	Resistance gene analog
RIT	Rennel Island Tall
RNA	Ribonucleic acid
RPB2	Second largest subunit of RNA polymerase II
RPS	Ribosomal protein sequence
SCAR	Sequence characterized amplified region
SCRI	Scottish Crop Research Institute
SERK	Somatic embryogenesis receptor kinase
SNP	Single nucleotide polymorphism
SOL	Solanaceae genomics network
SSR	Simple sequence repeat
SVM	Support vector machine
TAG	Triacylglycerol
WAT	West African Tall
WRI	Wrinkled

1 Introduction

1.1 Palms: Taxonomy and Uses

Palms belong to the viriplante group of higher plants under the group angiosperms, the closed-seed bearing plants. Angiosperm plants are further classified as monocot and dicots, based on number of seed leaves as being either one or two. Palms are classified as Arecaceae (palmae) botanical family, which is a single member of order Arecales (Table 1.1). Palms are an important group of plants for various human-end uses, next only to grasses and legumes. They are harvested from forests, cultivated as plantations or orchards, or used in landscaping indoor and outdoor environments. No exact statistical estimates specifiying the wide range of uses for the palms are available, making it difficult to list important palms and their specific valuable products. But, palms are used in a multitude of products such as food, wood, oil, fiber, and fuel. In fragile ecosystems such as tropical islands, coconut palm is life sustaining to islanders, who use it for food, oil, housing materials, and so on. Rattan is the major source of furniture, next in importance to bamboo. Palm wine and sugar are principal commodities traded. Coir, a fiber made from coconut, is used widely and traded around the world.

Palms grow in wide range of habitats, possess a wide range of traits, and serve many uses. Coconut, dates, arecanut, and oil palm are the four major palms commercially cultivated on large scale as orchards or plantations. Areas in which palms are grown, production, and major producers of these crops [1] are listed in Table 1.2. Coconut, an important perennial crop of the humid tropics, is a multipurpose crop that serves as a source of food, edible oil, refreshing drink, fiber (coir),

Table 1.1 Taxonomy of Cultivated Palms

Kingdom	Viriplantae
	Angisoperms
	Monocotyledons
Order	Arecales
Family	Arecaceae
Subfamilies	Calamoideae
	Nypoideae
	Coryphoideae
	Ceroxyloideae
	Phytelephantoideae
	Arecoideae
Tribes	13 tribes
Genera	Nearly 183
Species	Nearly 2450

Genomics of Cultivated Palms. DOI: 10.1016/B978-0-12-387736-9.00001-7

Table 1.2 Area of Production and Major Producers of Palms (Source: FAOSTAT, 2008)

Crop	Area (m ha)	Production (mt)	Yield (t/ha)	Major Producers
Arecanut	0.8	0.97	12.0	India (58%), China (13%), Bangladesh (10%)
Coconut	11.0	61	54	Philippines (25%), Indonesia (32%), India (18%)
Dates	1.2	7.0	55	Egypt (18%), Iran (14%), Saudi Arabia (14%), UAE (11%)
Oil palm	14.5	205	140	Indonesia (41%), Malaysia (40%)

fodder, and fuel. Arecanut is a source of supari and a mastigatory crop. Rattans are climbing palms used for making furniture; *Calamus*, *Daemonorops*, *Korthalsia*, and *Plectocomia* are major species of commercial significance. Oil palm is an important crop that offers a potential paradigm shift in the edible oil import and usage scenario. Date palm is a fruit crop, rich in dietary iron, growing in arid regions. Apart from these, palmyra (*Borassus flabellifer* L.) and pejibaye palms (*Bactris gasipaes*) are grown on large scale in India (Africa–Asia) and Cuba (America), respectively. Area under these palms is estimated as number of million palms rather than as hectares. Palms also serve as excellent ornamental plants and are either grown as indoor potted plants or in various outdoor landscapes, including avenue plantings. *Chrysalidocarpus lutescens* (erroneously referred to as areca palm or dypsis palm) is a good example of a most valued ornamental palm, whose seeds were once highly sought [2]. This palm can also remove indoor air pollutants such as formaldehyde [3]. Details on uses, diversity, and general aspects of tropical palms are described in current literature [4].

Palms first appear in the fossil record around 80–90 million years ago, during the late Cretaceous–Turonian period [5]. Arecanut palms are mentioned in ancient literature (1300 B.C.) of Sisu Mayana in India [6]. Palm leaves, especially those from the Thalipot palm or palmyra, were used as writing media during the fifth century B.C. [7] before the discovery of paper and print in Asian countries. Palms are mentioned in historical records and old religious and literature documents (3500 B.C.) such as the Bible [8]. The date palm was the first crop for which artificial pollination was attempted, by Babylonians and Assyrians, to improve the color of the fruit. Hand pollination of date palm has been practiced for centuries in Babylonia and is also well described in the Humorabi laws and in ancient Egyptian scripts [9].

1.2 Genetics Cytology and Genomics of Palms

Gene pool enrichment by indigenous and exotic germplasm, selection, and hybridization are the common breeding methods adopted in productivity enhancement. To obtain useful results, such efforts for perennial plants like palms require large amounts of land for experimentation, a long period of time, and adequate resources.

Despite these constraints, classical breeding tools were helpful in genetically enhancing palms for productivity, disease resistance, and so on. Some successful case studies are given below.

Coconut palms grow very tall and reach a height of 10–20m, making harvest feasible only with the help of laborers skilled in climbing. Dwarf trait is spontaneously noticed in coconut at relatively very low frequencies. Hybrids involving dwarf and tall forms of coconut offer promise in breeding for intermediate height, early bearing, and high yield. Lethal yellowing is a major phytoplasma disease killing millions of coconut palms in Latin American and Caribbean countries. Coconut germplasm available in these countries does not offer resistance to the malady. Cross-breeding Malayan dwarf coconut varieties from Malaysia, with local talls helped to produce more resistant strains [10].

Dura and pisifera are two common forms of oil palm. Dura has a thick-shelled kernel, whereas pisifera is a thin-shelled kernel type. Tenera is a high-yielding hybrid, a cross between dura and pisifera oil palms, differing in shell thickness (*Sh*+), a trait controlled by a single dominant gene [11]. Mutation breeding has been attempted in date palm to induce resistance to Bayoud disease [12]. Spontaneous mutants are utilized in coconut palm for jelly-like endosperm [13] controlled by a single recessive gene.

Chromosome numbers in most palms vary from $2n = 26$ to 36. Polyploidy is uncommon among the palms except the four species *Arenga caudata* ($2n = 4x = 64$), *Rhapis humilis* ($2n = 4x = 72$), *Jubaeopsis caffra* ($2n = 12x = 160$–200), and *Voanioala gerardii* (*V. gerardii* [Arecoideae]) with $2n = \sim 600$. Polyploidy is induced in oil palm, using chemical treatments of colchicine and oryzalin treatments [14]. Androgenic haploids are induced in coconut, using an *in vitro* culture of pollen [15]. DNA content, also known as *C*-value in a species, is measured using flow cytometry technique and is expressed in picograms. *C*-value is known for 104 species of palms as of date and varies from 0.25pg in cultivated date palm (*Phoenix dactlyfera*) to 39pg in *V. gerardii*, with a mean value of 4.04pg [16]. Karyological data of 56 palm taxa revealed that the chromosome number ranged from $2n = 36$ in ancestral species to $2n = 26$ as dysploid series in different subfamilies (Figure 1.1). Strong deviations in vegetative and floral morphology are accompanied by karyological differences [17]. Polyploidy is artificially induced in oil palm,

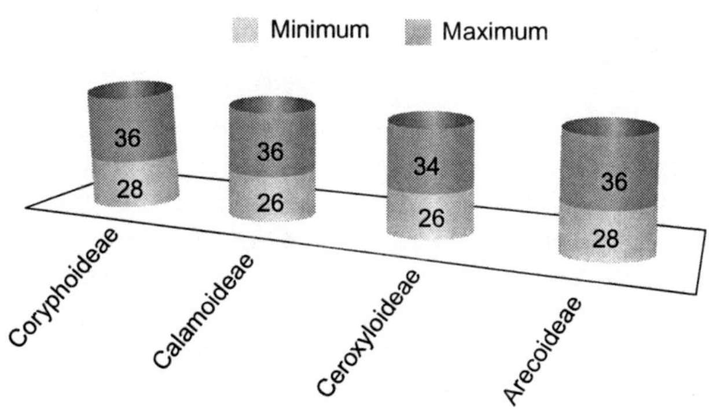

Figure 1.1 Range of chromosome numbers ($2n$) in subfamilies (Roser [17]).

Table 1.3 Botanical Name, Chromosome Number, Genome Size of Important Palms Compared to Completely Sequenced Genomes of Model Species

Crop	Botanical Name	Chromosomes	Genome Size (Million base pairs – Mbp)	Sequenced so Far (Mbp)
Coconut	*Cocos nucifera* L.	$2n = 32$	2950	0.07
Oil palm	*Elaeis guineensis* Jacq.	$2n = 32$	1950	10
Date palm	*Phoenix dactylifera* L.	$2n = 36$	540	360
Thale cress	*Arabidopsis thaliana* L.	$2n = 10$	150	150
Rice	*Oryza sativa* L.	$2n = 24$	400	400
Poplar	*Populus trichocarpa* L.	$2n = 38$	480	480

using colchicines and oryzalin [18]. Status on genome sequencing and the chromosome number of important palms vis-à-vis model plants are given in Table 1.3.

1.3 Molecular Phylogeny and Evolution

Palms are a monophyletic group of plants within the commelinid clade of monocotyledons [19]. Nearly 50 reports of phylogenetic studies of the arecaceae family as a whole or subfamilies have been published during the past two decades. DNA regions of partial sequences of gene or noncoding regions of four plastid [20], [maturaseK (matK), tRNA genes (trnL–trnF), ribosomal protein (rps16 intron), ribulose bisphosphate carboxylase large subunit (rbcL)], and two low-copy nuclear regions [21] phosphoribulokinase (PRK) and RBP2 (intron 2 of RNA polymerase) are used in the studies. Cloning and sequencing of these six regions was completed in more than 100 species of palms, and the sequences are available in the public domain. These sequences are mainly used to find the phylogenetic relationship and infer the evolution of palms. Recently, other nuclear DNA regions of sequences of WRKY genes [22] and conserved intron scanning primers [23] have also been are used in palms for the purpose.

1.4 Marker–Trait Associations

Now, many molecular and biochemical markers are available to analyze palm genomes. Linkage mapping programs have only been undertaken in oil palm and coconut, to locate the chromosomal regions of agronomic traits and molecular markers. Specific molecular markers associated with important economic traits are identified in different palms. Diseases and pests attack the plants and cause serious economic damage. Molecular markers associated with pest [24] resistance (coconut eriophyid mite, Figure 1.2) and disease [25] resistance (date palm Bayoud disease) are available and are useful in identifying promising seedlings for planting in pest- or disease-prone areas. Markers tightly linked to other yield traits are also identified in palms, such as those for shell thickness in oil palm [26]. Some of the palm species are dioecious, such as palmyra, date palm, and *Calamus* (rattan). Identification of sex

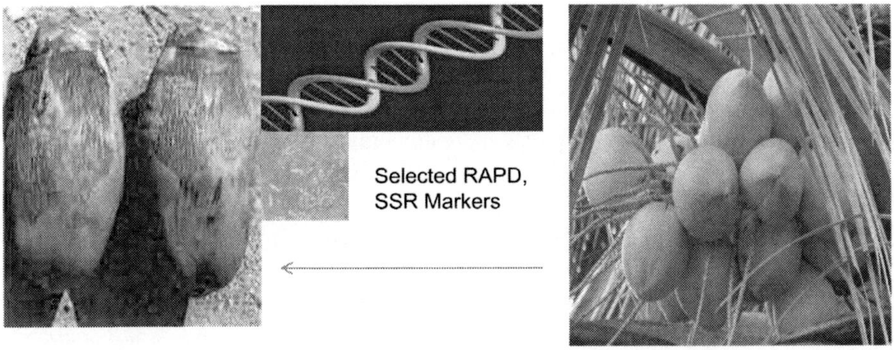

Figure 1.2 Marker–trait association in coconut mite resistance (Shalini et al. [24]).

Table 1.4 Status and Purpose of Genomic Resources in Palms

Crop	Transcriptomics	Proteomics	RAPD Markers	SSR Markers	QTL Mapping
Arecanut	Nil	Nil	Clonal fidelity	Diversity	Nil
Coconut	Endosperm tissues	Nil	Diversity, mite, LYD resistance	Diversity, mite, LYD resistance	Seedling traits, cuticular wax, nut yield, fruit components
Oil palm	Fatty acid biosynthesis	Nil	Diversity, shell thickness	Diversity, linkage	Yield, shell thickness, oil quality
Date palm	Nil	Seed germination, fungal colonization	Diversity	Diversity	Nil

in the seedlings is important, as the plants enter the reproductive stage after 10–12 years. Powerful DNA markers [27] can be used to identify the sex of the palms at the seedling stage. Undertaking marker–trait association studies on palms is the first step in developing genomic kits for use by farmers and nurserymen. In the following chapters, details are given for important traits in individual palms, such as sex, yield, quality, and resistance to biotic and abiotic stress factors. The status and purpose of genomic resources in different palms are listed in Table 1.4.

1.5 Molecular Cloning and Transgenics

Apart from the partial sequences of importance in phylogeny and evolution, molecular cloning of the principal genes governing important traits of palms has been attempted in palms. An important gene in coconut SERK (somatic embryogenesis receptor kinase) [28] that determines the plant's ability to regenerate via tissue culture is cloned and sequenced. The AP2 transcription factor gene that regulates meristematic and proliferating ability is sequenced in oil palm [29]. A promoter specific to endosperm tissue is isolated from coconut [30] as well as metal-inducible metallothionein promoter from oil palm [31]. Cloning of copia-like repetitive elements in coconut [32] led to the development of inverted sequence tagged repeat (ISTR) markers from the sequences. So far, among the palms, transgenic plants are reality only in oil palm [33].

1.6 Cross-Transferrable Microsatellite Markers in Palms

Microsatellite or simple sequence repeat (SSR) loci are powerful codominant markers to analyze genetic diversity and genome mapping. When these loci are conserved across the taxa, a given set of primers and the probability of amplifying the same (orthologous) gene in multiple species is great. Hence, such markers serve as a common tool to study multiple taxa such as the WRKY loci [22] in palms. Markers developed in model crop species such as rice are also useful for other monocots, including palms [34]. Microsatellite markers developed in a palm genus are also useful to other related palms. Cross-species amplification of microsatellite markers of a palm species [35–37] is reported in related species or genera (Table 1.5) of other palms.

1.7 Databases and Software

Advances in information technologies and sequencing technologies have made the job of analyzing and understanding the genomes and plants both cheap and quick. Online databases and genomic resources are available for aiding improvement of the four plant families (Table 1.6). A few resources are available for palms [40], *viz.*, coconut [41] and oil palm [42], and an expert system software, for ornamental palms [43]. There are also many Internet resources (Table 1.7) and a few cyberspace discussion groups (Table 1.8) available to help palm workers.

1.8 Constraints and Opportunities in Genomics of Palms

To a great extent, molecular biology tools aid in reducing the time, space, and resources necessary to enhance breeding efficiency. At present, molecular markers

Table 1.5 Cross-Species Validation of Microsatellite Markers in Palms

Originally Developed in	Cross-Species Amplification in	Reference
Oil palm	Coconut, *Jessenia batana*	[35]
Date palm	Oil palm, *Pritchardia, Astrocaryum*	[36]
Coconut	Rattans	[37]
Bactris gasipaes	Other species of *Bactris*	[38]
Date palm	Other species of Phoenix	[35,39]

are available to hasten crop improvement programs. Linkage mapping and QTL analysis programs help to decipher the markers closely associated with traits of interest such as shell thickness of oil palm [44] and cuticular wax [45] in coconut. Breeders are developing new markers to identify sex of plants and desirable traits

Table 1.6 Genomic Resources Developed as Portals Specific for Botanical Families

Family	Resource	Link	Major Plants Covered
Poaceae	Gramene	http://www.gramene.org/	Rice, sorghum, maize, wheat, and barley
Fabaceae	LIS	http://www.comparative-legumes.org/	Medicago, lotus, soybean, chickpea, groundnut, and pigeonpea beans
Rosaceae	GDR	http://www.bioinfo.wsu.edu/gdr/	Apple, pear, prunus, almond, apricot, cherry, peach, raspberry, rose, and strawberry
Solanaceae	SOL	http://solgenomics.net/	Tomato, potato, pepper, eggplant, petunia, tobacco, plus coffee, and snapdragon

Table 1.7 Internet Links of Societies, Organizations, and Networks for Palms

URL	Details of the Organization/Network/Society
http://www.pacsoa.org.au	PACSOA Palm and Cycad Societies of Australia
http://www.plantapalm.com/	PACSOF Palm and Cycad Societies of Florida
http://www.plantapalm.com/vpe/vpe_index.htm	Virtual palm encyclopedia
http://www.neiker.net/link2palm/Coco/DefCOCO.htm	Link2Palm of Neiker Institute
http://www.inbar.int/	International Network for Bamboo and Rattan
http://tropgenedb.cirad.fr/	Tropgenedb
http://www.inibap.org/cogent/	The International Coconut Genetic Resources Network
http://www.kew.org/	Royal Botanic Garden, Kew, UK
http://www.apccsec.org/	The Asian and Pacific Coconut Community
http://www.palmweb.org/	Palms of the World Online (Palmweb)
http://www.scanpalm.no/english.html	Palms in Scandinavia
http://www.palmsociety.org.uk/	The European Palm Society
http://www.icar.org.in	Indian Council of Agricultural Research
http://www.cirad.fr	CIRAD Montpellier France
http://www.mpob.gov.my	Malaysian Palm Oil Board
http://www.cri.lk	Coconut Research Institute Sri Lanka
http://www.pca.da.gov.ph	Philippine Coconut Authority
http://www.cpcri.gov.in	Central Plantation Crops Research Institute India
http://www.iopri.org	Indonesian Oil Palm Research Institute
http://www.aciar.gov.au	Coconut and its pathogens
http://www.qatar-weill.cornell.edu/research/datepalmGenome/download.html	Date palm genome sequencing project

Table 1.8 Discussion Groups for Palms

Crop	Discussion Group
Coconut	http://groups.google.com/group/coconut
Date palm	http://www.datepalm.org/
Oil palm	http://www.agrisoft-systems.de/AS_OP_Forum_01.htm
Rattans	http://groups.inbar.int:8888/forums/list.do
Palms of Scandinavia	http://www.scanpalm.no/english.html

such as dwarf habit and resistance to disease or drought. After wet-lab verification experiments, these markers can be used to offer a set of kits that could identify the seedlings of palms possessing the desirable traits.

Recently, novel markers of industrial relevance offer promise, such as identification of palm oil-contaminated edible oil samples, using unique sequences of oil palm genes MT-3 [46]. Such markers must be validated, stringently screened, and developed as user-friendly kits. Genomic studies also provide tools for basic research in understanding the slower evolution in genes in palms, especially *Adh* (alcohol dehydrogenase), than in grasses [47], and also in nurturing the classical systematic research of palms [48]. A set of AFLP markers was employed to analyze the sympatric evolution in Howea palms in oceanic islands [49]. The complete genome of date palm is available now [50] and that of oil palm is expected soon, making the palms an interesting subject for genomic researchers.

References

1. FAOSTAT. (2010). http://faostat.fao.org/default.aspx
2. Johnson DV. (ed.) 1996. *Palms: their conservation and sustained utilization.* Status survey and conservation action plan. Gland, Switzerland: IUCN The World Conservation Union.
3. Matsumoto H and Yamaguchi M. (2007). Experimental study on the effect of foliage plants on removing indoor air contaminants. *Proceedings of Clima 2007 WellBeing indoors*, pp. 1–8.
4. Johnson DV. (2010). Non wood forest products no. 10\Rev. 1. *Tropical palms.* Food and Agricultural Organisation, Rome, Italy, 242 p.
5. Harley MM. A summary of fossil records for Arecaceae. *Bot. J. Linn. Soc.* 2006;**151**:39–67.
6. Bhat PSI, Rao KSN. On the antiquity of arecanut. *Arecanut J.* 1962;**13**:13–21.
7. Jones R. The origins of the Malay manuscript tradition. In: Grijns CD, Robson SO, eds. *Cultural Contact and Textual Interpretation.* Holland: Foris Publications; 1986.
8. Janick J. Fruits of the Bibles. *Hort. Sci.* 2007;**42**:1072–1076.
9. Al-Bakr A. The date palm. Baghdad, Iraq: Al-Ani Press; 1972.
10. Been BO. Observations on field resistance to lethal yellowing in coconut varieties and hybrids in Jamaica. *Oléagineux.* 1981;**36**:9–11.
11. Beinaert A and Vanderweyen R. (1941). Contribution à l'étude génétique et biométrique des variétés d'Elaeis guineensis Jacq. Publications de l'institut national pour l'étude agronomique du Congo Belge, série scientifique, 27 pp.

12. Jain SM. Recent advances in date palm tissue culture and mutagenesis. *Acta Hort.* 2007;**736**:205–211.
13. Zuñiga LC. The probable inheritance of the Makapuno character of coconut. *The Philippine agriculturist.* 1953;**36**:402–413.
14. Al-Ani B, Zaid A, Shabana H. On the status of chromosomes of the date palm (*Phoenix dactylifera* L.). *Acta Hort.* 2010;**882**:253–268.
15. Perera PIP, Hocher V, Verdeil JL, Bandupriya HDD, Yakandawala DMD, Weerakoon LK. Androgenic potential in coconut (*Cocos nucifera* L.). *Plant Cell Tissue Organ Cult.* 2008;**92**:293–302.
16. Bennett MD, Leitch IJ. Nuclear DNA amounts in angiosperms: 583 new estimates. *Ann. Bot. (Lond.).* 1997;**80**:169–196.
17. Roser M. Pathways of karyological differentiation in palms (Arecaceae). *Plant. Syst. Evol.* 1994;**189**:83–122.
18. Madon M, Clyde MM, Hashim H, Yusuf MY, Mat H, Saratha S. Polyploidy induction of oil palm through colchicine and oryzalin treatments. *J. Oil Palm Res.* 2005;**17**:110–123.
19. Baker WJ, Savolainen V, Asumussen-Lange CB, Chase MW, Dransfield J, Forest J, et al. Complete generic level phylogenetic analyses of palms (Arecaceae) with comparisons of supertree and supermatrix approaches. *Syst. Biol.* 2009;**58**:240–259.
20. Asumussen CB, Dransfield J, Deckmann V, Barfod AS, Pintaud J-C, Baker WJ. A new subfamily classification of the palm family (Arecaceae): evidence from plastid DNA phylogeny. *Bot. J. Linn. Soc.* 2005;**151**:15–38.
21. Roncat J, Zona S, Lewis CE. Molecular phylogenetic studies of Caribbean palms (Arecaceae) and their relationship to biogeography and conservation. *Bot. Rev.* 2008;**74**:78–102.
22. Meerow AW, Noblick L, Borrone JW, Couvreur LP, Mauro-Herrera M, Hahn WJ, et al. Phylogenetic analysis of seven WRKY genes across the palm subtribe Attaleinae (Arecaceae) identifies Syagrus as sister group of the coconut. *PLoS ONE.* 2009;4(10):e:7353.
23. Bacon CD, Feltus FA, Paterson AH, Bailey CD. Novel nuclear intron-spanning primers for Arecaceae evolutionary biology. *Mol. Ecol. Resour.* 2007;**8**:211–214.
24. Shalini KV, Manjunantha S, Lebrun P, Berger A, Baudouin L, Pirany N. Identification of molecular markers associated with mite resistance in coconut (*Cocos nucifera* L.). *Genome.* 2007;**50**:35–42.
25. Quenzar B, Trifi M, Bouachrine B, Hartmann C, Marrakchi M, Benslimane A, et al. A mitochondrial molecular marker of resistance to Bayoud disease in date palm. *Theor. Appl. Genet.* 2001;**103**:366–370.
26. Moretzsohn MC, Nunes CDM, Ferreira ME, Grattapaglia D. RAPD linkage mapping of the shell thickness locus in oil palm (*Elaeis guineensis* Jacq.). *Theor. Appl. Genet.* 2000;**100**:63–70.
27. Li M, Li F, Yang F, Yin G, Gan S. A male-specific SCAR marker in *Calamus simplicifolius*, a dioecious rattan species endemic to China. *Mol. Breed.* 2010;**25**:549–551.
28. Perez-Nunez MT, Souza R, Saenz L, Chan JL, Zuniga-Aguilar JJ, Oropeza C. Detection of a SERK-like gene in coconut and analysis of its expression during the formation of embryogenic callus and somatic embryos. *Plant Cell Rep.* 2009;**28**:11–19.
29. Morcillo F, Gallard A, Pillot M, Jouannic S, Bertossi FA, Collin M, et al. EgAP2-1, an AINTEGUMENTA-like (AIL) gene expressed in meristematic and proliferating tissues of embryos in oil palm. *Planta.* 2007;**226**:1353–1362.
30. Xu L, Ye R, Zheng Y, Wang Z, Zhou P, Lin Y, et al. Isolation of the endosperm-specific LPAAT gene promoter from coconut (*Cocos nucifera* L.) and its functional analysis in transgenic rice plants. *Plant Cell Rep.* 2010;**29**:1061–1068.

31. Omidvar V, Abdullah SNA, Izadfard A, Ho CL, Mahmood M. The oil palm metallothionein promoter contains a novel AGTTAGG motif conferring its fruit-specific expression and is inducible by abiotic factors. *Planta*. 2010;**232**:925–936.
32. Rohde W, Salamani F, Ashburner R, Randles JW. An EcoRI repetitive sequence family of the coconut palm (*Cocos nucifera* L.) shows sequence homology to copia-like elements. *J. Genet. Breed.* 1992;**46**:391–394.
33. Lee M, Yeun L, Abdullah R. Expression of *Bacillus thuringiensis* insecticidal protein gene in transgenic oil palm. *Electron. J. Biotechnol.* 2006;**9**:117–126.
34. Lohithaswa HC, Feltus FA, Singh HP, Bacon CD, Bailey CD, Paterson AH. Leveraging the rice genome sequence for monocot comparative and translational genomics. *Theor. Appl. Genet.* 2007;**115**:237–243.
35. Ting NC, Noorhariza MZ, Rozana R, Low ET, Ithnin M, Cheah SC, et al. SSR mining in oil palm EST database: application in oil palm germplasm diversity studies. *J. Genet.* 2010;**89**:135–145.
36. Billotte N, Marseillac N, Brottier P, Noyer JL, Jacquemoud-Collet JP, Moreau C, et al. Nuclear microsatellite markers for the date palm (*Phoenix dactylifera* L.): characterization and utility across the genus Phoenix and in other palm genera. *Mol. Ecol. Notes.* 2004;**4**:256–258.
37. Rao NM, Ramesha BT, Ravikanth G, Ganeshaiah KN, Umashankar R. Cross-species amplification of coconut microsatellite markers in Rattans. *Silvae Genet.* 2007:282–286.
38. Martenez AK, Gaitin-Solis E, Duque MC, Bernal R, Thome J. Microsatellite loci in *Bactris gasipaes* (Arecaceae): their isolation and characterization. *Mol. Ecol. Notes.* 2002;**2**:408–410.
39. Akkak A, Scariot V, Marinoni T, Boccacci P, Beltramo C, Botta R. Development and evaluation of microsatellite markers in *Phoenix dactylifera* L. and their transferability to other Phoenix species. *Biol. Plant.* 2009;**53**:164–166.
40. Ruiz M, Rouard M, Raboin LM, Lartaud M, Lagoda P, Courtois B. TropGENE-DB, a multi-tropical crop information system. *Nucleic Acids Res.* 2004;**32**:D364–D367.
41. Rajagopal V, Manimekalai R, Devakumar K, Rajesh MK, Karun A, Niral V, et al. A database for coconut crop improvement. *Bioinformation.* 2005;**1**:75–77.
42. Riju A, Chandrasekhar A, Arunachalam V. Mining for single nucleotide polymorphisms and insertions/deletions in expressed sequence tag libraries of oil palm. *Bioinformation.* 2007;**2**:128–131.
43. Beck H, Jones P, Watson D, Zazueta F. An expert database system for ornamental plants. *Agric. Syst.* 1989;**31**:111–126.
44. Billotte N, Marseillac N, Risterucci AM, Adon B, Brottier P, Baurens FC, et al. Microsatellite-based high density linkage map in oil palm (*Elaeis guineensis* Jacq.). *Theor. Appl. Genet.* 2005;**110**:754–765.
45. Riedel M, Riederer M, Becker D, Herran A, Kullaya A, Arana-López G, et al. Cuticular wax composition in *Cocos nucifera* L.: physicochemical analysis of wax components and mapping of their QTLs onto the coconut molecular linkage map. *Tree Genet. Genomes.* 2009;**5**:53–69.
46. Zhang L, Wu G, Wu Y, Cao Y, Xiao L, Lu C. The gene MT3-B can differentiate palm oil from other oil samples. *J. Agric. Food Chem.* 2009;**57**:7227–7232.
47. Morton BR, Brandon T, Gaut S, Clegg MT. Evolution of alcohol dehydrogenase genes in the palm and grass families. *Proc. Natl. Acad. Sci. U.S.A.* 1996;**93**:11735–11739.
48. Dransfield J, Uhl NW, Asmussen NCB, Baker WJ, Harley MM, Lewis CE. Genera palmarum: the evolution and classification of palms. Kew, UK: Royal Botanic Gardens; 2008.

49. Savolainen V, Anstett M, Lexer C, Hutton I, Clarkson JJ, Norup MV, et al. Sympatric speciation in palms on an oceanic island. *Nature*. 2006;**441**:210–213.
50. Al-Dous EK, George B, Al-Mahmoud ME, Al-Jaber MY, Wang H, Salameh YM, et al. De novo genome sequencing and comparative genomics of date palm (*Phoenix dactylifera*). *Nat. Biotech*. 2011;**29**:521–527.

2 Coconut

2.1 Introduction

Coconut (*Cocos nucifera* L.) is an important multipurpose perennial crop of the tropics. Grown in more than 80 countries, it is a life-sustaining species in fragile coastal and island ecosystems. Indonesia, Philippines, India, and Sri Lanka are the major coconut-producing countries. Coconut belongs to the monotypic genus *Cocos*, under the subtribe Butiinae; and tribe Cocoeae, subfamily Arecoideae. It is a diploid species ($2n = 2x = 32$) with a large genome size of 2.15 Gb nucleotide pairs [1]. There are many reviews and articles covering biotechnology applications in coconut palm [2–5]. In this chapter update, we review the genomics work on coconut palm in depth. Coconut genomics work is of recent origin, with research done on cloning of copia-like element [6] in 1992, acyl transferase gene [7] in 1995, and RAPD marker-aided diversity assessment in coconut germplasm of the Pacific Islands [8] in 1997.

2.2 Markers for Assessment of Diversity

Molecular markers (RAPD [8–11]), AFLP [12,13], RFLP [14], ISTR [15], ISSR [16], SSR [17–25], WRKY genic SNP [26] and WRKY genic SSR [27]), and biochemical markers (polyphenols [28], enzymes [29,30], and proteins [31]) were employed to understand the diversity in coconut germplasm. Details of the materials used, primers tried, and the conclusions are given in Tables 2.1a–2.1d.

Tall and dwarf are the two main habitual forms available in the coconut palm [32]. Tall cultivars are predominantly heterogeneous due to their cross-pollinated nature. Tall cultivars possess late-bearing (7–10 years) stems with long internodes, produce large fruits, and grow very tall (20–30 m). Dwarf cultivars grow to a height of 10–15 m and occur at very low frequency in nature. They are early bearing (5–6 years), predominantly self-pollinated, and thereby homogenous. They possess thin stems with short, compact internodes and produce small fruits with thin copra [33]. There are three types of dwarf forms in coconut. First is the javanica, intermediate types with semi-tall habit and medium-sized fruits; the second type, nana, are real dwarf types with dwarf habit and small fruits. Hybrids involving tall and dwarf forms gave early-bearing, high-yielding intermediate forms with medium-long internodes [34,35]. Lethal yellowing is a serious disease of coconut in Latin American and Caribbean countries. Introduction of coconut germplasm in Ref. [36] led to observation of the resistance to the disease in Malayan dwarfs. Hybrids between dwarf and tall cultivars for high yield and resistance to lethal yellowing are the two important achievements in coconut genetics.

Genomics of Cultivated Palms. DOI: 10.1016/B978-0-12-387736-9.00002-9

Table 2.1a Molecular and Biochemical Markers in Coconut: RAPD Markers

Primers Used	Germplasm/Materials	Conclusion	Reference
14 primers mostly OPA series	14 populations of Pacific Islands	Pattern of continuous variation but 2 southern groups with domesticated type; northeastern group with wild type fruits	[8]
80 primers (Operon AH, OP, A, C, M series) 16 polymorphic	28 palms of 3 yellow dwarfs (Malayan [MYD], Andaman [AYD], and Kulasekaram [KYD])	Greater diversity within accessions than between accessions; MYD is putative origin of other 2 dwarfs; KYD16 individual is distinct	[9]
8 primers mostly from OPA, B, C series	81 palms of 20 accessions (15 from India, 5 from Indonesia, Philippines, Malaysia)	Narrow diversity in Indian accessions, tall genotypes high heterogeneity; dwarfs did not cluster together	[10]
45 primers (Operon C, M, E, and BE series)	33 accessions from all coconut-growing continents	Five informative primers: OPE6, OPC13, OPF14, OPM2, and OPM17; high diversity in New Caledonian accessions	[11]

Niu Leka, or Fiji dwarf, is the third type of dwarf; it is cross-pollinated and has large fruits but compact stems with short nodes. The genetic nature of the Fiji dwarf is elaborate [20] along with unique microsatellite allele possessed by the cultivar. A standard three- or four-letter abbreviation delineates coconut cultivars by their geographical origin and habit. The last letter in the abbreviation denotes the habit of the cultivar. For example, RIT refers to the tall cultivar of Rennell Island in the South Pacific, known as Rennell Island Tall. Coconut populations are broadly classified in five groups, based on their geographical origin, as South Asia, Southeast Asia, African, South Pacific, and Latin American and Caribbean. Foliar polyphenols [28] and their components are found to be a reliable indicator of geographical origin of the coconut. Coconut populations on a coast are similar to each other compared to those on other coasts, although they may be located on the same continent. West African coconut populations on the Atlantic coast [29] are genetically proximal to those in Mexico on the Atlantic coast. The highest level of genetic diversity with the polymorphic microsatellite alleles [19] and RAPD assay [8] is seen in Pacific Island talls. Most of the molecular marker studies confirmed the heterogeneity in tall coconut populations and their differentiation from homogenous dwarf populations [9,11,13,22]. Three different SSR marker kits are available and being used for

Table 2.1b Molecular and Biochemical Markers in Coconut: RFLP, ISTR, ISSR, and AFLP Markers

Methods Used	Germplasm/Materials	Conclusion	Reference
AFLP: 8 primer combinations of EcoRI and MseI	42 genotypes from 8 of typica, 4 of nana, and 2 of aurantiaca from Sri Lanka	Aurantiaca group genetically similar to dwarf	[12]
AFLP 12 *Mse*I and *Pst*I primer combinations—1106 bands, 303 polymorphic SSR 39 SSRs; 37 polymorphic CNZ, CN1, CN11, CN2 series	31 individuals from 10 tall and 4 dwarf populations from across the globe	Two of the four dwarf populations were homozygous at all 37 loci, due to autogamous nature; Niu Leka dwarf displayed heterozygosity; South Pacific and SouthEast Asia region—high diversity; 3 SSR loci (CNZ46, CN2A5, CN11E6) discrimated most populations The East African populations had higher heterozygosities than those from West Africa; Tonga and Fiji populations had unique alleles from rest of South Pacific	[13]
RFLP—enzymes EcorI, BglII, Ssti, and rice cDNA probes	100 individuals of 10 tall, 7 dwarf accessions, and 2 hybrids	40 polymorphic bands; 2 major groups of coconut accessions: (Far east which also includes all dwarfs and South Pacific) Comoro tall intermediate between the two	[14]
ISTR	New marker from copia-like elements cloned from coconut	20–100 Loci can be detected; transferrable marker to other palms, plants, and animals	[15]
ISSR—19 primers, 199 ISSR markers, 154 polymorphic	33 accessions (29 talls, 2 intermediates, and 3 dwarfs) of entire globe	Least similarity between Nicobar tall and chowghat orange dwarf. Southeast Asia, South Asia, and South Pacific formed separate groups	[16]

Table 2.1c Molecular and Biochemical Markers in Coconut: SSRs (17–25)

Markers Used	Country	Germplasm/ Materials	Conclusion	Reference
8 CAC SSRs (2, 3, 4, 6, 8, 10, 11, 13)	Global	130 (75 tall and 55 dwarf) of 94 ecotypes	8 SSRs could discriminate 116 out of 130 analyzed; 5 main groups found with either talls or dwarfs	[18]
12 CAC SSRs	Global	130 individuals of 51 talls and 43 dwarfs	2 groups (first group with talls of southeast Asia, Pacific, west coast of Panama, and all dwarfs; second group with other talls of south Asia, Africa, and Indian Ocean coast of Thailand)	[19]
15 CAC SSRs	Florida, USA	56 individuals of 5 populations; 54 individuals of 9 cultivars, 1 hybrid	Identification of population specific alleles for genotypic identification, especially for Fiji dwarf	[20]
14 CnCir SSRs	India	102 palms of 10 landraces (8 talls; 2 dwarfs) in 3 locations of South India	Expected heterozygosity highest in the 2 tall populations of Pallikara but least in 2 dwarfs of Vayalar. Landraces clustered by geographical locations and breeding behavior	[21]
41 SSRs CnZ series; 8 SSRs in detail	Philippines	Initially 40 accessions; 10 tall and 10 dwarf cultivars in detail	Dwarf cultivars grouped separately from talls and displayed less genetic diversity	[22]
26 SSRs from CAC, CNZ, and CnCir	Hainan Islands, China	45 individuals of 10 accessions, hybrids from 6 locations	Hybrids formed distinct and homogenous groups; high genetic diversity; gene exchange between individuals was limited	[24]
13 CnCir SSRs	Dominican Republic	13 accessions from 9 locations of Criollo local tall cultivar; 2 hybrids	Criollo genetically close to Atlantic tall of Mexico and belong to Indo Atlantic group	[25]

Table 2.1d Molecular and Biochemical Markers in Coconut: Biochemical and WRKY Genic SNPs, SSRs

Methods used	Germplasm/ Materials	Conclusion	Reference
WRKY-SNPs: 10 markers—SSCP analysis	15 individuals of 6 cultivars	CnWRKY-13 showed 3 alleles; 5 WRKY-SNPs are in linkage disequilibrium (LD), possibly by inbreeding of cultivars	[26]
WRKY-SNPs: 13 markers—SSCP analysis	110 individuals as used (20)	CnWRKY-01 showed 3 alleles; one is unique to Red Spicatas and Fiji dwarfs; a second is unique only to Fiji dwarfs; all other cultivars are homozygous for the third allele; only slight differences between WRKY SNP and CAC SSRs, and other marker systems; Green Nino and Malayan dwarf are grouped together	[27]
Peroxidase, endopeptidase, glucose 6 phosphate dehydrogenase (G6PD)	21 Mexican talls, 6 other talls, and 2 hybrids	Dwarfs separately grouped from tall but similar to Pacific talls; low heterozygosity endogamy genetic drift in differentiated accessions of Pacific coast from Gulf of Mexican coast	[29]
10 enzymes—14 loci	3 palms each in 30 cultivars (19 talls, 7 dwarfs, 4 hybrids)	Highest heterozygosity in G6PG and least in SOD (super oxide dismutase)	[30]
Leaf protein, peroxidiase, endopeptidase	2 talls, 2 dwarfs, and 2 hybrids	4 distinct genotypes identified: one each for tall cultivar, another for dwarf cultivar, and last for both hybrids	[31]

coconut. One is a set developed as CnZ series in the Philippines [22]; another set is the CAC series developed in Scotland [17–20]; and a third is a set of SSRs developed in France as the CnCir series [23]. Primers and sequences of these SSR loci are available at Genbank database, and map positions are at the Link2palm website (http://www.neiker.net/link2palm/Coco/DefCOCO.htm). Microsatellite markers are currently used widely in the characterization of coconut germplasm belonging to different countries such as China [24], India [21], Sri Lanka [17], Philippines [22], Dominican Republic [25], and USA (Florida) [20].

2.3 Maker-Trait Associations, QTL Mapping on Coconut

Markers associated with important traits can increase the breeding efficiency and save time in a perennial crop like coconut. Resistance to eriophyid mite [37], lethal yellowing disease [38], and lethal disease [39] are associated with selected alleles of SSR or RAPD markers (Table 2.2). These markers, after further validation and conversion as SCAR markers and kit or chip, serve as an efficient tool for coconut breeding. QTL mapping of important traits such as early flowering [40], yield [41], fruit components [43], and composition of cuticular wax [44] are identified so far (Table 2.2). Strong QTLs and saturation of coconut linkage map with an additional set of markers can increase the breeding efficiency of coconut for the traits of interest. AFLP markers are the favorable choice in QTL mapping experiments [40–43] as they generate a large number of scorable bands. All other marker systems such as ISTRs [40], ISSRs [41], RAPDs [37,38,40], RFLPs [42], and SSRs [37,39,41–43] were also useful in trait association/QTL mapping in coconut. Details of the knowledge available on mapping of genes and markers with traits are as given in Table 2.2.

2.4 Genome Biology of Coconut Endosperm and Fatty Acid Biosynthesis

Endosperm is the most important economic part of the coconut palm and provides a refreshing health drink during the immature stage and edible grade oil during the mature stage. Little information is available on the coconut endosperm, from the experiments on cDNA library [44], X-ray diffraction studies of cocosin [45], gene [7] and promoter [46] cloning of LPAAT, miRNA chip [47] on mature and immature endosperm, free nucleotides [48], and biochemistry [50] of a mutant endosperm. Sequencing of the transcriptome of the 8- and 15-month-old endosperm revealed that 41% of genes are homologous to those having known gene function. Expression of oleosin and globulin genes (similar to oil palm ones) is specific to 15-month-old pulp and not in 8-month-old pulp. One unique gene, KIAA 1239 protein, 322 amino acids long, which is otherwise known only in mammalian systems like humans, is overexpressed at the endosperm development stage of coconut [44]. Coconut endosperm contains a storage protein known as cocosin. Cocosin was purified, crystallized,

Table 2.2 Marker-Trait Associations and QTL Mapping in Coconut

Trait(s)	Mapping Populations/ Cultivars Used	Markers Analyzed	Conclusion	Reference
Eriophyid mite resistance	40 coconut trees of South India, differing mite damage	32 CnCir SSRs, of 160 Operon RAPDs, 7 RAPDs (OPE4, 6, 18, OPF2, OPG7, OPP15, 16) in detail	3 SSRs and 2 RAPD markers associated	[37]
Lethal yellowing disease resistance	20 individuals in each of resistant (Atlantic Tall (AT), Malayan Yellow Dwarf (MYD), susceptible West African Tall (WAT) cultivars)	80 primers of Operon series A, B, AB, and AL	82 RAPD bands differentiated DNA pools; 5 markers found associated (OPB4-570, OPA11-990, OPB11-1140, OPAL3-1160, and OPAL7-350)	[38]
Lethal disease resistance	6–8 individuals of tolerant (Vanuatu Tall [VTT], Sri Lankan green Dwarf [SGD]), and susceptible West African Tall (WAT)	12 CnCir series of SSR loci	10 specific alleles associated with tolerance identified	[39]
Seedling traits	52 progenies of Malayan Yellow Dwarf (MYD20) × a Laguna Tall (LAGT07)	24 ISTR, 5 ISSR, and 44 AFLP primer, 54 bands of 28 RAPDs	382 markers mapped; 6 QTLs for early germination; genome length 2226 cM to 1266 cM for LAGT and MYD respectively; 4–32 markers/chromosome	[40]
Yield: number of bunches and fruits	(CRD) Cameroon Red Dwarf (P04383-4394) × Rennell Island Tall (RIT) PO2664	62 AFLP combinations; CnCir CNZ SSRs (13, 22), RFLPs (14)	227 markers mapped; 9 QTLs for number of fruits	[41]
Fruit components		230 (RFLP, SSR, and AFLP)	290 markers (80 Cn SSRs, 8 Eg SSRs), 6 RFLPs, 204 AFLPs mapped; 52 putative QTLs for 11 traits	[42]
Epicuticular wax and its components	East African Tall (1011) × Rennel Island Tall (0710)	241 AFLP markers, 64 coconut, and 22 oil palm SSRs	46 QTLs and 1026 markers and a gene ACD01 desaturase gene mapped	[43]

and analyzed by X-ray diffraction. Crystals of cocosin are rhombohedral with space group R3 with 92.829, 92.829, and 215.290Å dimension [45]. The composition of a high proportion of medium-chain fatty acids is an important feature of coconut oil. LPAAT (1-acyl-Sn-glycerol-31 acyl transferase) lysophosphatidyl acyl transferase is an important enzyme that catalyzes conversion of lysophosphatidic acid (LPA) to phosphatidic acid (PA). Activity of the LPAAT enzyme shows a preference for medium-chain length fatty acyl–coenzyme A substrates. The gene controlling the LPAAT enzyme is cloned [7] and sequenced, and is 1.325kb in size (U29657) with an anticipated protein of 308 amino acids (Table 2.3). Promoter of the LPAAT gene of 1.2kb size has recently been sequenced, and contains TATA box, CAAT box, Skn-1 motif, W box, AGGAA, and TCCC motifs [46].

MicroRNAs are small regulatory RNA molecules 20–22 nucleotides long. They negatively regulate the expression of target by post-transcriptional gene silencing. About 179 microRNAs (miR) were investigated at China by authors of [47] using microRNA chip, of which 95 are expressed in the immature and 176 are expressed in the mature stage of endosperm. Twenty-three miRs were upregulated and 9 were down-regulated in mature endosperm. Computational analysis identified the 32 target genes for these coconut miRs. Three miRs (Ath-miR398a, b, and bna-miR397a miRs) displayed unique expression only in immature endosperm. Laccase was identified as a target gene for the unique miR [47]. Free nucleotides present in the coconut endosperm were analyzed and were found to be natural miRNA:mRNA hybrid molecules [48].

Makapuno, also spelled as *macapnuo* and called variously in different locations ghee coconut, sap, Coco Gra, kopyor, and Dikripol (review: 4), is a delicacy in coconut. It is due to a mutation [49] at a single mendelian gene (mm), where the endosperm becomes buttery and jelly-like rather than like a common solid endosperm (MM). Makapuno embryos cannot germinate in nature due to this deleterious allele causing the endosperm-hindering embryo germination. But the plants having heterozygous locus for the makapuno mutation (Mm) produce normal (MM/Mm) and makapuno (mm) fruits based on the gamete received from normal (M) and Makapuno (m) pollen cells, respectively. Biochemical analysis revealed the macapuno fruits to have 8300-fold-reduced activity of the enzyme alpha galactosidase compared to normal fruits [50]. Free nuclei of the coconut endosperm have the potential to be a bioreactor for expressing recombinant proteins [51].

2.5 Coconut Tissue Culture-Related Genes

Tissue culture of oil and date palms is successful and commercially practiced. But the technique is still being standardized in coconut, with nonrepeatable results. A review of coconut biotechnologies, especially in *in vitro* propagation [52], gives some background of the past work. Sequencing of a few genes controlling the somatic embryogenesis, such as cyclic dependant kinase (CDKA-)[53], knotted-like homeobox (KNOX-)[54], and somatic embryogenesis receptor kinase (SERK-)[55], has resulted in important information (Table 2.3). The details of the genes, their role,

Table 2.3 Genes and Promoters Cloned from Coconut

Gene	Role	Accession No.	Conclusion	Reference
LPAAT	Catalyzes conversion of lysophosphatidic acid (LPA) to phosphatidic acid (PA)	U29657	1.325 kb size anticipated protein of 308 amino acids; promoter of LPAAT gene (1.2 kb) with Skn-1 motif, W box, AGGAA, and TCCC motifs	[37] [46]
CDKA	Cellular control—studied for the role in somatic embryogenesis during embryogenic callus formation	GQ924086	882 bp long (294 amino acids) CnCDKA gene similar to Picea expression steadily increased during embryogenic competence	[53]
KNOX	Knotted homeobox transcription factor induced by giberellic acid (GA_3) studied here	CnKNOX1 (GU937114) CnKNOX2 (GU937115) with KNOX and ELK domains	CnKNOX1, only in meristematic tissues 1002 bp long (333 amino acids) CnKNOX2 (54 bp long 317 amino acids) in both meristematic and nonmeristamtic	[54]
SERK	Somatic embryogenesis receptor kinase	AY791293 with SPP domain, leucine zipper domain and 5 LRR domains	2.24 kb long 1.89 kb ORF (629 amino acids) resembled rice SERK and rice benzothiadiazole–induced SERK	[55]
WRKY	Transcription factor with WRKY domain; WRKY families of 2, 6, 7, 16, and 19	(WRKY2—FJ956959 to 64; WRKY6—FJ957104 to 09; WRKY7—FJ957178 to 83; WRKY12—FJ957250-01;WRKY16—FJ957319-21; WRKY19—FJ957389 to 94;WRKY21—FJ957030 to 35)	Using a degenerate primer pair, WRKY gene regions were cloned and sequenced from 6 coconut populations	[66]
MAPK	Chitosan-induced mitogen-activated protein kinase	AJ555489-91 with TEY domain	Addition of chitosan activates a 46 kDa MAPK-like protein	[67]

and features are listed in Table 2.3. KNOX genes of coconut resembled that of oil palm, whereas CDKA genes of coconut resembled the tree Picea or rice or maize. SERK genes of coconut resembled rice benzothiadiazole-induced SERK genes. The role of benzothiadiazole in enhancing tissue culture is inferred from the study. We attempted a similar search of coconut SERK genes using NCBI BLAST, but we could not get any match with known SERK genes for the first 70–80 base pairs of the sequence. However, this region matched with either a hypothetical protein or a whole genome shotgun sequence, which is worth exploring.

2.6 Structural Biology of Hirsutellin

Eriophyid mite, an important pest of coconut in LAC countries, later spread to many Asian and African countries. The pest has become a serious threat to coconut cultivation except in the Pacific region. Chemical control methods are neither eco-friendly nor easy to practice. Biological control and resistance breeding are the only options left. Markers for resistance are identified [37] and need to be converted to reliable SCAR markers and kits after validation on a large number of samples. The biological control agent of fungal pathogen *Hirsutella thompsonii* [56] offers promise. Hirsutellin A (HtA) is an extracellular protein produced by the invertebrate, resembling a ribotoxin similar to sarcin. The structural biology of the protein [59–61] reveals many interesting observations. Initially, HtA toxin was identified and was not proved to have a role in pathogenicity [57]. Later studies confirmed the toxicity and led to identification of the complete gene sequence [58] and corresponding protein [59]. NMR studies indicated a characteristic feature of a alpha+beta fold with a high positive charge at surface [60]. Asp40 in the smallest ribotoxin (HtA) is crucial in providing electrostatic interactions and biological activity [61]. Details of the studies on the toxin and its structural biology are given in Table 2.4.

2.7 Road Map for Coconut Genomics

A database resource was developed for all coconut researchers, covering genomics, pathology, microbiology, and genetics [62]. Another database of coconut [63] is a useful resource with markers and map position data for the coconut genomic research community. Microsatellites developed for coconut are also useful in rattans [64], oil palm [65], and other palms [66]. WRKY sequences recently obtained from coconut [66] and protein kinase sequences [67] are important resources for development of gene-specific markers in palms for tagging disease resistance and phylogenetics studies. Coconut genomics is still in the take-off stage and can reach similar heights to those of oil and date palm by harnessing comparative genomics. Molecular marker resources and kit are available for the characterization of coconut germplasm. The knowledge available on marker-trait association and/or QTLs for seedling traits, yield, fruit component traits, epicuticular wax components, and resistance to lethal yellowing disease and eriophyid mite needs further validation. These

Table 2.4 Structural Biology of Hirsutellin A (HtA) Toxin for Eriophyid Mite of Coconut

Aim	Methods and Materials	Conclusion	Reference
Detection of HtA toxin	162 strains of infected mites collected worldwide and anlayzed for HtA gene PCR amplification	100 of them gave PCR amplification for HtA gene; HtA toxin needed for either survival or pathogenicity	[57]
Characterization of HtA gene	Cloned and sequenced HtA gene (U86836)	Ribosomal protein (RIP) predicted to be 164 amino acids long	[58]
Solution structure of the toxin	NMR (nuclear magnetic resonance) technique	Similar to other ribotoxins a alpha + beta fold; highly positive charge at surface	[59]
Structural characterization	*In vitro* synthesis, western blot, spectroscopic analysis, phospholipid assay, cytotoxicity	A fungal ribonuclease of 130 amino acids long (~17 kDa) specifically inactivates ribosomes proven by cytotoxicity assay	[60]
Crucial residue(s) for activity	Comparison of HtA and a-sarcin three-dimensional structures	HtA smallest ribotoxin; Asp40 equivalent to Sarcin Tyr48 important for electrostatic interaction	[61]

markers then require conversion to SCAR markers and kit or chip, to serve as an efficient tool for coconut breeding. Sequences of a few transcription factors (knotted gene, WRKY) and genes governing somatic embryogenesis (SERK, CDK) and oil quality (acyl transferase), as well as promoter specific to endosperm tissue from coconut, have been deciphered.

Genomic knowledge of tissue culture from oil palm and coconut must be converted to useful protocols for enhanced tissue culture tools for coconut. Gene mapping of other important traits of coconut palm, such as dwarf habit, resistance to other diseases, and abiotic stress (drought, cold), is a future genomics research priority. The structural biology of hirsutellin, toxin for the biological control of the coconut mite, is known and should be analyzed further for appropriate use.

References

1. Roser M, Johnson AT, Hanson L. Nuclear DNA amounts in palms (Areaceae). *Botanica Acta.* 1997;**110**:79–89.
2. Batugal P, Bourdeix R, Baudouin L. Coconut breeding. In: Jain SM, Priyadarshini PM, eds. *Breeding Plantation Tree Crops: Tropical Tree Species*. Berlin: Springer Science + Business Media LLC; 2009, p. 327–375.

3. Perera L, Perera ACNS, Bandaranayake CK, Harries HC. Coconut oil crops. Vollman J. Rajcan I, editors. Handbook of Plant Breeding, 4. Berlin: Springer Science + Business Media, LLC; 2009, p. 369–396,10.1007/978-0-387-77594-4_12.
4. Arunachalam, V. and Rajesh, M.K. (2008). Breeding of coconut palm. CAB Rev.: Perspect. *Agric., Vet. Sci., Nutr. Nat. Resour.* **3**, 1–12.
5. Harries HC, Baudouin L, Cardena R. Floating, boating and introgression: molecular techniques and the ancestry of coconut palm populations on Pacific Islands. *Ethnobotany Res. Appl.* 2004;**2**:37–53.
6. Rohde W, Salamani F, Ashburner R, Randles JW. An EcoRI repetitive sequence family of the coconut palm (Cocos nucifera L.) shows sequence homology to copia-like elements. *J. Genet. Breed.* 1992;**46**:391–394.
7. Knutzon DS, Lardizabal KD, Nelsen JS, Bleibaum J, Davies HM, Metz JC. Cloning of a Coconut Endosperm cDNA Encoding a 1 -Acyl-sn-Clycerol-3-Phosphate Acyltransferase that Accepts Medium-Chain-Length Substrates. *Plant Physiol.* 1995;**109**:999–1006.
8. Ashburner GR, Thompson WK, Halloran GM. RAPD analysis of South Pacific coconut palm populations. *Crop Sci.* 1997;**37**:992–997.
9. Paul R, George J, Rajesh MK, Jerard BA, Niral V. Genetic diversity in yellow dwarf populations of coconut assessed using RAPD markers. *Indian J. Hort.* 2008;**65**:60–64.
10. Upadhyay A, Jayadev K, Manimekalai R, Parthasarathy VA. Genetic relationship and diversity in Indian coconut accessions based on RAPD markers. *Sci.Hort.* 2004;**99**:353–362.
11. Manimekalai1 R, Nagarajan P. Interrelationships among coconut (Cocos nucifera L.) accessions using RAPD technique. *Genet Resour. Crop Evol.* 2006;**53**:1137–1144.
12. Perera L, Russell JR, Provan J, McNicol JW, Powell W. Evaluating genetic relationships between indigenous coconut (Cocos nucifera L.) accessions from Sri Lanka by means of AFLP profiling. *Theor. Appl. Genet.* 1998;**96**:545–550.
13. Teulat B, Aldam C, Trehin R, Lebrun P, Barker JH, Arnold GM, et al. An analysis of genetic diversity in coconut (Cocos nucifera L.) populations from across the geographic range using sequence-tagged microsatellites (SSRs) and AFLPs. *Theor. Appl. Genet.* 2000;**100**:764–771.
14. Lebrun P, N'Cho YP, Seguin M, Grivet L, Baudouin L. Genetic diversity in coconut (Cocos nucifera L.) revealed by restriction fragment length polymorphism (RFLP) markers. *Euphytica.* 1998;**101**:103–108.
15. Rohde W, Kullaya A, Rodriguez MJB, Ritter E. Genetic analysis of Cocos nucifera L. by PCR amplification of spacer sequences separating a subset of copia-like EcoRI repetitive elements. *J. Genet. Breed.* 1995;**49**:179–186.
16. Manimekalai R, Nagarajan P. Assessing genetic relationships among coconut (Cocos nucifera L.) accessions using inter simple sequence repeat markers. *Sci. Hort.* 2006;**108**:49–54,10.1016/j.scienta.2006.01.006.
17. Perera L, Russell JR, Provan J, Powell W. Identification and characterization of microsatellite loci in coconut (*Cocos nucifera* L.) and the analysis of coconut populations in Sri Lanka. *Mol. Ecol.* 1999;**8**:344–346.
18. Perera L, Russell JR, Provan J, Powell W. Use of microsatellite DNA markers to investigate the level of genetic diversity and population genetic structure of coconut (Cocos nucifera L). *Genome.* 2000;**43**:15–21.
19. Perera L, Russell JR, Provan J, Powell W. Studying the genetic relationships among coconut varieties/populations using microsatellite markers. *Euphytica.* 2003;**132**: 121–128.

20. Meerow AW, Wisser RJ, Brown JS, Kuhn DN, Schnell RJ, Broschat TK. Analysis of genetic diversity and population structure within Florida coconut (Cocos nucifera L.) using microsatellite DNA, with special emphasis on the Fiji Dwarf cultivar. *Theor. Appl. Genet*. 2003;**106**:715–726.
21. Rajesh MK, Arunachalam V, Nagarajan P, Lebrun P, Samsudeen K, Thamban C. Genetic survey of 10 Indian coconut landraces by simple sequence repeats (SSRs). *Scientia Horticulturae*. 2008;**118**(4):282–287, 10.1016/j.scienta.2008.06.017.
22. Rivera R, Edwards KJ, Barker JHA, Arnold GM, Ayad G, Hodgkin T, et al. Isolation and characterization of polymorphic microsatellites in Cocos nucifera L. *Genome*. 1999;**42**:668–675.
23. Baudouin L, Lebrun P. The development of a microsatellite kit and dedicated software use with coconuts. *Burotrop. Bull*. 2002;**17**:16–20.
24. Liu X, Tang H, Dongdong Li, D, Hou L. Genetic diversity of coconut cultivars in China by microsatellite (SSR) markers. *Mol. Plant Breed*. 2011;**2**:83–91.
25. Martinez RT, Baudouin L, Berger A, Dollet M. Characterization of the genetic diversity of the Tall coconut (Cocos nucifera L.) in the Dominican Republic using microsatellite (SSR) markers. *Tree Genet. Genomes*. 2010;**6**:73–81, 10.1007/s11295-009-0229-6.
26. Mauro HM, Meerow AW, Borrone JW, Kuhn DN, Schnell RJ. Ten informative markers developed from WRKY sequences in coconut (*Cocos nucifera*). *Mol. Ecol. Notes*. 2006;**6**:904–906.
27. Mauro HM, Meerow AW, Borrone JW, Kuhn DN, Schnell RJ. Usefulness of WRKY gene-derived markers for assessing genetic population structure: an example with florida coconut cultivars. *Sci. Hort*. 2007;**115**:19–26.
28. Jay M, Bourdeix P, Potier F, Sanslaville E. Initial results from the study of polymorphism of coconut leaf phenols. *Oleagineux*. 1989;**44**:158–161.
29. Zizumbo-Villarreal D, Cardena-Lopez R, Pinero D. Diversity and phylogenetic analysis in Cocos nucifera L.in Mexico. *Genet. Resour. Crop Evol*. 2002;**49**:237–245.
30. Geethalakshmi P, Parthasarathy VA, Niral V. Genetic diversity among coconut genotypes using isozymes. *Asian J. Plant Sci*. 2005;**4**:678–683.
31. Cardena R, Oropeza C, Zizumbo-Villareal D. Leaf proteins as markers useful in the genetic improvement of coconut palms. *Euphytica*. 1998;**102**:81–86.
32. Narayana GV, John CM. Varieties and forms of the coconut. *Madras Agric. J*. 1949;**36**:349–366.
33. Sugimura Y, Itano M, Salud CD, Otsuji K, Yamaguchi H. Biometric analysis on diversity of coconut palm: cultivar classification by botanical and agronomical traits. *Euphytica*. 1997;**98**:29–35,10.1023/A:1003053128120.
34. Kumaran PM, Arunachalam V, Nampoothiri KUK, Pillai RV, Ratnambal MJ, Niral V, et al. Performance of coconut hybrids in rainfed conditions. *J. Plan. Crops*. 2006;**34**:11–14.
35. Bourdeix R. Coconut selection and breeding. In: Ohler JG, ed. *Modern Coconut Management*. Rome: FAO; 1999. p. 458.
36. Been BO. Observations on field resistance to lethal yellowing in coconut varieties and hybrids in Jamaica. *Oleagineux*. 1981;**36**:9–12.
37. Shalini KV, Manjunantha S, Lebrun P, Berger A, Baudouin L, Pirany N. Identification of molecular markers associated with mite resistance in coconut (*Cocos nucifera* L.). *Genome*. 2007;**50**:35–42.
38. Cardena R, Ashburner GR, Oropeza C. Identification of RAPDs associated with resistance to lethal yellowing of the coconut (Cocos nucifera L.) palm. *Sci. Hort*. 2003;**98**:257–263.

39. Noël KKJ, Edmond KK, Louïs KJ, Lebrun P, Dery SK, Sangare A. Microsatellite gene diversity in coconut (Cocos nucifera L.) accessions resistants to lethal yellowing disease. *Afr. J. Biotech.* 2007;**6**:341–347.
40. Herran A, Estioko L, Becker D, Rodriguez MJB, Rhode W, Ritter E. Linkage mapping and QTL analysis in coconut (Coconut nucifera L.). *Theor. Appl. Genet.* 2000;**101**: 292–300.
41. Lebrun P, Baudouin L, Bourdeix R, Konan JL, Barker JHA, Aldam C, et al. Construction of a linkage map of the Rennell Island Tall coconut type (Cocos nucifera L.) and QTL analysis for yield characters. *Genome.* 2001;**44**:962–970.
42. Baudouin L, Lebrun P, Konan JL, Ritter E, Berger A, Billotte N. QTL analysis of fruit components in the progeny of a Rennell Island Tall coconut (Cocos nucifera L.) individual. *Theor. Appl. Genet.* 2006;**112**:258–268.
43. Riedel M, Riederer M, Becker D, Herran A, Kullaya A, Arana-López G, et al. Cuticular wax composition in Cocos nucifera L.: physicochemical analysis of wax components and mapping of their QTLs onto the coconut molecular linkage map. *Tree Genet.Genomes.* 2009;**5**:53–69.
44. Li D, Fan YM. Construction and characterization of a cDNA Library from the pulp of coconut (Cocos nucifera L.). *Agric. Sci. China.* 2008;**7**:1071–1076.
45. Balasundaram D, Sugadev R, Ponnuswamy MN. Purification and crystallization of coconut globulin cocosin from cocos nucifera. *Biochim. Biophys. Acta.* 2002;**1601**:121–122.
46. Xu L, Ye R, Zheng Y, Wang Z, Zhou P, Lin Y, Li D. Isolation of the endosperm-specific LPAAT gene promoter from coconut (Cocos nucifera L.) and its functional analysis in transgenic rice plants. *Plant Cell Rep.* 2010;**29**:1061–1068.
47. Li D, Zheng Y, Wan L, Zhu X, Wang Z. Differentially expressed microRNAs during solid endosperm development in coconut (Cocos nucifera L.). *Sci. Hort.* 2009;**122**:666–669.
48. Cueno ME, Laude RP. Molecular characterisation of free nucleic acids in coconut water. *Philippine J. Sci.* 2010;**139**:35–41.
49. Torres J. Some notes on makapuno coconut and its inheritance. *Philippines J. Agric.* 1937;**8**:27–37.
50. Mujer CV, Ramirez DA, Mendoza EMT. Alpha-Dgalactosidase deficiency in coconut endosperm: its possible pleiotropic effects in makapuno. *Phytochemistry.* 1984;**23**:893–894.
51. Bustamante JO. New biotechnological applications of coconuts. *Electron. J. Biotech.* 2004;7.
52. Oropeza C, Verdeil JL, Ashburner GR, Cardena R, Santamaria JM. Current Advances in Coconut Biotechnology. Dordrecht: Kluwer Academic; 1999, 439 p.
53. Montero-Cortes M, Rodrıguez-Paredes F, Burgeff C, Perez-Nunez T, Cordova I, Oropeza C, et al. (2010). Characterisation of a cyclin-dependent kinase (CDKA) gene expressed during somatic embryogenesis of coconut palm. Plant Cell Tissue Organ Cult. 102, 251–258. 10.1007/s11240-010-9714-8.
54. Montero-Cortes M, Sanenz L, Cordova I, Quiroz A, Verdeil J-L, Oropeza C. GA3 stimulates the formation and germination of somatic embryos and the expression of a KNOTTED-like homeobox gene of Cocos nucifera (L.). *Plant Cell Rep.* 2010;**29**:1049–1059.
55. Perez-Nunez MT, Souza R, Saenz L, Chan JL, Zuniga-Aguilar JJ, Oropeza C. Detection of a SERK-like gene in coconut and analysis of its expression during the formation of embryogenic callus and somatic embryos. *Plant Cell Rep.* 2009;**28**:11–19.
56. Fisher FE. Entomogenous fungi attacking scale insects and rust mites on citrus in Florida. *J. Econ. Entomol.* 1950;**43**:305–309.

57. Maimala S, Tartar A, Boucias DG, Chandrapatyab A. Detection of the toxin Hirsutellin A from Hirsutella thompsonii. *J. Invertebr. Pathol.* 2002;**80**:112–126.
58. Boucias DG, Farmerie WG, Pendland JC. Cloning and sequencing of cDNA of the insecticidal toxin Hirsutellin A. *J. Invertebr. Pathol.* 1998;**72**:258–261.
59. Viegas A, Herrero-Galán E, Oñaderra M, Macedo AL, Bruix M. Solution structure of hirsutellin A—new insights into the active site and interacting interfaces of ribotoxins. *FEBS J.* 2009;**276**:2381–2390.
60. Herrero-Galán E, Lacadena J, Martínez-del Pozo AM, Boucias DG, Olmo N, Oñaderra1 M, et al. The insecticidal protein hirsutellin A from the mite fungal pathogen Hirsutella thompsonii is a ribotoxin. *Proteins: Structure, Function, and Bioinformatics.* 2008;**72**:217–228.
61. Herrero-Galán E, García-Ortega L, Lacadena J, Martínez-Del-Pozo A, Olmo N, Gavilanes JG, et al. (2011). Implication of an Asp residue in the ribonucleolytic activity of hirsutellin A reveals new electrostatic interactions at the active site of ribotoxins. Biochimie (in press).
62. Rajagopal V, Manimekalai R, Devakumar K, Rajesh MK, Karun A, Niral V, et al. A database for coconut crop improvement. *Bioinformation.* 2005;**1**:75–77.
63. Ruiz M, Rouard M, Raboin LM, Lartaud M, Lagoda P, Courtois B. TropGENE-DB, a multi-tropical crop information system. *Nucleic Acids Res.* 2004;**32**:D364–D367.
64. Rao NM, Ramesha BT, Ravikanth G, Ganeshaiah KN, Umashankar R. Cross-species amplification of coconut microsatellite markers in Rattans. *Silvae Genet.* 2007:282–286.
65. Billotte N, Marseillac N, Risterucci AM, Adon B, Brottier P, Baurens FC, et al. Microsatellite-based high density linkage map in oil palm (*Elaeis guineensis* Jacq.). *Theor. Appl. Genet.* 2005;**110**:754–765.
66. Meerow AW, Noblick L, Borrone JW, Couvreur LP, Mauro-Herrera M, Hahn WJ, et al. Phylogenetic analysis of seven WRKY genes across the palm subtribe Attaleinae (Arecaceae) identifies Syagrus as sister group of the coconut. *PLoS ONE.* 2009;**4**(10): e: 7353.
67. Lizama-Uc G, Estrada-Mota IA, Caamal-Chan MG, Souza-Perera R, Oropeza-Salin C, Islas-Flores I, et al. Chitosan activates a MAP-kinase pathway and modifies abundance of a defense-related transcripts in calli of Cocos nucifera L. *Physiol.Mol. Plant Pathol.* 2007;**70**:130–141.

3 Oil Palm

3.1 Introduction

Oil palm is the second largest source of edible oil, next only to soybean. It contributes approximately one-fifth of the world's production of oils and fats, and belongs to the genus *Elaeis*. There are two important species in the genus *Elaeis*, *E. guineensis* (African oil palm) and *E. oleifera* (American oil palm). African oil palm is commercially cultivated and a high-yielding species. American oil palm is a low-yielding species, but the oil quality and composition is better than that of the African species. Both species of oil palm are diploid ($2n = 2x = 32$) and are crossable. African oil palm trees are classified into three groups based on fruit morphology. The dura (D) has a thick shell (but without the ring of fibers) enclosing the endosperm or kernel. The pisifera (P) is usually female sterile and has no shell, but it has a ring of fibers enclosing the kernel, whereas the tenera (T) has a thin shell as well as a ring of fibers enclosing the kernel. Tenera hybrids boosted the oil yield by 30% in oil palm, and the research in oil palm genetics boosted the yield fourfold from 1947 to 1997 [1]. Many studies and articles cover biotechnology applications in oil palm [2–6]. In this chapter, we review the genomics work on oil palm in depth. Tissue culture is a successful way to propagate the oil palm crop. But an abnormality of sterile inflorescence is noticed in 3–5% of the tissue-cultured palms known as mantled. Transgenic plants are a reality in oil palm [7], and many foreign genes are successfully incorporated in oil palm such as cry1Ac for insect resistance [8]; GUS, a reporter gene; and polyhydroxy butyrate [9] for use as bioreactor. Molecular markers, genes, QTL maps, and transcriptome resources have been studied in oil palm [2] more than in any other palm.

3.2 Markers, QTLS, Omics of Mesocarp, and Shell Thickness

The fruit of the oil palm (*Elaeis guineensis* Jacq.) is a drupe whose thick, fleshy mesocarp is exceptionally rich in oil content (80% of dry weight), and the species hence is the highest oil-yielding crop of the world [6]. Shell thickness is the most important trait in oil palm, which differs between the dura and the pisifera, the two common fruit forms of oil palm. Dura (Plate 3.1) is a form with a thick-shelled kernel, whereas pisifera (Plate 3.2) is a thin-shelled kernel type. Tenera is a high-yielding hybrid between the dura ($Sh{+}Sh{+}$) and the pisifera ($Sh{-}Sh{-}$) forms of oil palms, differing in the trait of shell thickness ($Sh{+}$), controlled by a single locus with two alleles in a codominant fashion in the tenera form ($Sh{+}Sh{-}$) [10,11]. Pisifera is a mutation in oil palm associated with a failure of lignin and fiber

Genomics of Cultivated Palms. DOI: 10.1016/B978-0-12-387736-9.00003-0

Plate 3.1 Dura-the thick shelled form of oil palm.

Plate 3.2 Pisifera-the thin shelled form of oil palm.

formation in the region where the shell is formed. Pisifera palms are usually female sterile and are difficult to propagate by seeds.

The endocarp cells of oil palm fruit become lignified to a hard shell, while the mesocarp tissue remains fibrous during shell formation. A transition zone made up of fibrous units was also visible beneath the shell. The role of ferulic acid and the three enzymes (phenylalanine ammonia lyase [PAL], cinnamyl alcohol-NADPH-dehydrogenase [CAD], and peroxidase [POD]) is crucial in lignin synthesis [12]. Hence, the quantification of activity of these lignin biosynthetic enzymes is also used to discriminate the seedlings as dura, pisifera, and tenera [12,13].

3.2.1 Markers and Genes for Shell Thickness

Molecular markers associated with shell thickness can aid in identification of fruit form at the seedling stage of oil palm. They were identified using RAPD [14,15], AFLP [16], RFLP [17], and SSR [18,19] marker systems. RAPD bands specific to pisifera (P12-600, P10-700, P6-650, P19-800), dura (P7-700, P10-1000), and tenera (P28-1100) [14] are identified. These markers need further validation and conversion as SCAR markers. Three specific operon RAPD markers, OPY20–1180, OPR11–1282, and OPT19–1046, are found to be associated with the locus of shell thickness [15]. The estimated map distances from *Sh+* to OPR11–1282 and to OPT19–1046 were 17.5 cM and 23.9 cM, respectively, and were identified as linked on both sides of the *Sh+* locus on linkage group 4 [15]. These two RAPDs were able to predict shell thickness with an accuracy of nearly 97%. An AFLP marker E-AGG/M-CAA132 was mapped at 4.7 cM from the *Sh* locus [16]. Another RFLP marker (pOPgSP1282) closest to the locus was located 9.8 cM from the *Sh+* locus [17]. About 75 out of 100 plants were heterozygous, with primer mEgCIR 008 having a $(GA)_{17}$ repeat at an annealing temperature of 52 °C, with eight alleles varying in size from 105 to 150 bp; and the tenera hybrid can be distinguished from dura and pisifera parents [18]. Another SSR loci, mEgCIR1772, also serves the purpose, but only in specific families [19]. Currently, the 14 molecular markers (9 RAPD primers, 2 AFLP loci, 1 RFLP loci, and 2 SSR loci) from these independent studies together need further validation. Previous studies on the role of ferulic acid and enzymes PAL, CAD, and POD and other lignin biosynthetic enzymes [12,13] must be clarified, using current molecular biology, sequencing, and bioinformatics technologies along with fine mapping of the *Sh+* locus. Map-based cloning, sequencing, and *in silico* analysis of the corresponding fourth chromosomal segment have the potential of making great strides in boosting the oil yield of oil palm.

3.2.2 Omics of Oil Palm Mesocarp Biology

Lipid-rich, fleshy mesocarp of oil palm and sugar-rich mesocarp of date palm were compared using transcriptomics and metabolomics [20]. Regulatory mechanisms in oil palm fruit ripening and mesocarp content were studied by pyrosequencing of transcriptome and analysis of fatty acid (FA) and triacylglycerol (TA) assembly pathways [21]. Surprisingly, both these independent studies identify a single seed oil transcription factor, WRINKLED1, as an overexpressed (57-fold) transcript in oil palm [20] mesocarp. The ratio of cell wall invertase to sucrose synthase is 3.5 times greater in oil palm transcripts than *Arabidopsis* [20]. The activity of AGPase and starch synthase enzymes increases during ripening of oil palm fruits [20]. NAC-domain proteins and MADs-box genes, in particular SEP-like, AG-like, and GLO-like, were found to be involved during maturation and ripening of oil palm mesocarp, especially the AGL2/SEPALLATA subfamily [21]. Lipase class 3 family [22] and ethylene receptor genes [21,22] are overexpressed specifically in the mesocarp of oil palm fruits. Maturation proteins (PM3) are the genes specific to kernel tissues, and S-ribonuclease-binding protein and fibrillin were leaf and mesocarp tissue-specific genes [23]. Differentially expressed genes are found in processes associated with oil palm nut maturation, such as the synthesis of medium-chain saturated fatty acids and phytic acid, nut development, and stress/defense responses [24]. A sesquiterpene synthase gene unique to 12- to 20-week-old mesocarp of *E. oleifera*, but absent in *E. guineensis* and other tissues of both species [25], requires further analysis to establish an association with oil quality of the American oil palm.

Promoters are regions upstream of the coding sequences of the genes and play a crucial role in tissue-specific and condition-specific expressions. A mesocarp tissue-specific promoter of the mesocarp-specific metallothionein-like gene promoter (also known as MSP1; accession no: EU499363) corresponding to the oil palm MT3-A gene has been cloned and sequenced for use in the genetic engineering of oil palm [26]. This promoter is tested using the *uidA* reporter gene in bacterial [27] systems (*E. coli, Agrobacterium tumefaciens*). The promoter is metal inducible in the presence of metals such as copper due to a copper-inducible motif GTAC in the sequence. The 986 bp fragment also has TATA box and ethylene-responsive elements [27] and fruit tissue-specific AGTTAGG motif [28]. The promoter is induced by the presence of abscisic acid (ABA) and copper [28]. Further, secrets of oil biosynthesis are reviewed and summarized [29] with special reference to oil palm.

3.3 Transcriptomics and Genes of Somatic Embryogenesis and Mantled Disorder

Transcriptome analysis of callusing and somatic embryogenesis of oil palm identified a glutathione S-transferase as a highly expressed gene in nonembryogenic callus [30,34]. Unigenes analysis of the transcriptome indicated 648 nonredundant simple sequence repeats and 211 putative full-length open reading frames [30]. Egpk1, EgHOX1, EgPER1, and OPSC10 are the four highly expressed genes in embryogenic cultures of oil palm [31]. EgLSD (a putative lignostilbene-a,b-dioxygenase) and EgER6 (a putative ethylene-responsive gene) were the auxin-responsive genes in the suspension cultures of oil palm [32]. A putative serine-threonine kinase (DQ 267436) was also found expressed in tissue cultures of oil palm [33]. Recently, 928 annotated unigenes were identified in the transcripts during the somatic embryogenesis [34]. In this study, a transcript for a guanine nucleotide-exchange protein GEP2 was found only in the initiation library (GH635901–GH637767). The majority of the defense-related transcripts in the library were coded as either methallothionein-like or peroxidase-related proteins. Other ESTs with protein destination and storage functions are related to the ubiquitin/26S proteosome pathway [34]. In the embryoid cDNA library of oil palm, 325 unigenes (DW 247764–DW 248770) were identified in different tissues: germinA in 15-week-old kernel, Myb1trancription factor in spear leaves, and LEA (late embryogenesis abundant) genes in seedling roots [35].

Multiplication of oil palm by tissue culture via somatic embryogenesis produces 3–5% of off-type plants with a homeotically modified flower structure known as *mantled*. In these plants, fertile or sterile androecium is transformed into carpel-like structures, thereby lowering oil yields in mantled palms. Mantled and normal inflorescences were analyzed using molecular biology, and two genes associated with the difference were discovered as *EgFB1* and *EgRING*. These two genes are down-regulated in the mantled inflorescences and known to encode for proteolytic signaling proteins [36]. MET, CMT, and DRM are the methyl transferases (EU 117216–EU117218), and DNA methylation differences were noticed in mantled and normal inflorescence of oil palm [37]. While analyzing the 14,537 expressed sequences of (GenBank accession numbers EL680967–EL695503) from different tissues of oil palm [38], a few genes involved in different aspects of flower development were also identified. The majority of the genes related to flowering, *EgSB, EgSEP CONSTANS-like, AGAMOUS-like (AGL)2, AGL20, LFY-like, SQUAMOSA, SQUAMOSA-binding protein (SBP)*, and so on, are unique in oil palm and are useful in studying mantled disorder [38]. Flower and etiolated seedling tissue-specific myo-inositol phosphate synthase-like genes were found in oil palm while studying the differential gene expression [39].

cDNA libraries generated from the tissues of male and female inflorescences, shoot apices, and zygotic embryos (BM402088, BM402089, CN599371–CN601781) were analyzed to understand mantled disorder [40]. Glycine-rich RNA-binding proteins were seen as a common and abundant gene in all five tissues. A BURP domain-containing protein and a PVR3-like protein were specific to apical tissues. Storage proteins (7S globulins and oleosins) were specific to zygotic embryos [40]. A comparative study on transcripts of somatic and zygotic embryos revealed the accumulation of globulin 7S during mid to late zygotic embryogenesis, but the absence of protease activity and transcripts of cysteine proteinase (*CPR*) gene. Transcripts of three cysteine proteinase genes of the papain family were detected as early as the second day of *in vitro* germination [41]. An Eg 707 gene of abscisic acid (ABA) biosynthesis was found to display a high degree of expression only in somatic embryogenesis in *in vitro* tissues of oil palm [42]. A recent review of 15 years of research on mantled disorder provides further details [43].

3.4 Molecular Markers for QTL Mapping and Diversity Analysis in Oil Palm

Linkage analysis and QTL mapping experiments in oil palm used a wide range of molecular markers (AFLP [16,44,45], cDNA RFLP [46], SSR [47,48], and RFLP [50]) or a combination of AFLP, AFLP, and SSR [49]. The following is the list of traits mapped on oil palm genome so far (Table 3.1).

1. Yield and its components (fruit weight and ratio of shell and kernel to fruit [50]), oil-to-wet mesocarp ratio [46], and shell thickness [14–19].
2. Quality (fatty acid composition and iodine value) of oil [49].
3. *Fruit color* [45]: Virescence (vir) is a term that describes the change of fruit color in which mature fruits are emerald green and turn bright orange on ripening. It is a single dominant gene over Nig [11].
4. Phenotypic traits of stem and leaf [47,48].

Molecular markers were also employed to understand the diversity of oil palm germplasm native to different African and American countries (Table 3.2). AFLP [51,52], RAPD [53–57], RFLP [58–60], SSR [61–66], isozymes [67], and SNP [68–69] have potential to understand the diversity and also serve as markers to aid in breeding oil palm. Among the RAPD primers, OPT06 [55] and UBC731 [56]

Table 3.1 Linkage Analysis and QTL Mapping Experiments in Oil Palm

Marker System(s)	Mapping Population	Traits and Markers Mapped, Coverage of Markers/Conclusion	Reference
48 AFLP primers 288 markers	87 progenies of dura (ENL48) × pisifera (ML161)	Mapping of AFLP markers in dura and pisifera progenies; 127 markers in 22 linkage groups. Genetic distance of 555 cM in dura and 885 cM in pisisfera	[44]
688 AFLP loci, 256 oil palm, 21 coconut SSR loci	Tenera × dura (LM2T × DA10D) and map integrated with data of DA10D	Shell thickness on 77 cM terminal segment of LG 4; marker density of 1.8 cM covers the length of LG 59–192 cM total coverage 1743 cM	[16]
321 cDNA RFLP probes from 14 enzymes	Selfed single (T128) tenera palm	Oil-to-wet mesocarp ratio on LG1; coverage 693 cM	[46]
10 AFLP loci	Nigrescence × virescence progenies	E-ACT-MCTA-254 and E-ACT-M-CAT-142,356 flank virescence on LG3	[45]
509 SSRs, candidate genes	Tenera × dura (LM2T × DA10D); coconut (EAT0707 × PRD)	371 SSRs mapped on oil palm 216 on coconut map; traits: stem height, length of leaf, shell thickness	[47]
390 oil palm and 21 coconut SSRs	Multiparents 4 tenera, Deli (DA10D), LM2T, LM718T, dura (LM269D)	76 QTLs associated with 24 quantitative phenotypic traits; QTLs common across families identified	[48]
413 AFLP loci, 289 RFLP, 33 SSR	Interspecific hybrid (P1026 × CT128) (E.g × E.o) *E. guinensis* × *E. oleifera*	Oil quality by fatty acid composition, iodine value; 199 AFLPs, 38 RFLPs, 15 SSRs mapped major QTLs associated and identified their location	[49]
103 RFLPs mainly of PstI enzyme	Self-pollinated progenies of A137/30, Dami progenies (1327,1182) E37/08 × A137/30, 1 E.O and 14 E.g accessions	Shell thickness	[17]
153 RFLPs	Selfed progenies of individual (A137/30)	Yield and phenotypic traits markers specific to each trait identified fruit weight: 2 cM—PE95; kernel to fruit ratio: 1 cM—SP1342; shell to fruit ratio: 3 cM—SP298	[50]

Table 3.2 AFLP, RAPD, RFLP, Isozyme, and SSR Markers for Genetic Diversity in Oil Palm

Markers	Germplasm or Somaclonal Variation or Breeding	Conclusion	Reference
AFLP			
233 bands of three primers	Native to Nigeria, Madagascar, Democratic Republic of Congo (Congo), and Gambia	88% polymorphism	[51]
AFLP 5 loci Isozymes 4 systems	Deli, Congo, and Cameroon/Côte d'Ivoire	Crosses among African subpopulations to be included in reciprocal recurrent selection rather than African × Deli crosses.	[52]
RAPD			
96 primers; 12 specific	175 Caiaue accessions of Amazon of Brazil + 17 African accessions	Moderate diversity in caiaue group; high diversity in Negro and Madeiro accessions; sampling of many individuals rather than populations for covering diversity at Amazon.	[53]
20 primers; 9 polymorphic	Many accessions of oil-palm germplasm collected from Africa	No individual palm or population-specific products were observed. Greatest diversity was seen in Zaire population 5 and the least in Zaire population 2.	[54]
160 primers; 7 specific	151 plants from different areas in southern Thailand	OPT06 primer is useful for tenera identification.	[55]
RAPD, SSR markers	Oil palm germplasm	UBC 731 is useful in cultivar identification.	[56]
387 primers; 259 amplified 73 polymorphic	Mother palms, regenerants, mantled disordered ones	RAPD is useful for somaclonal variation, but not for mantled disorder.	[57]
RFLP			
5 enzymes; 4 cDNA probes	359 accessions from 11 African countries + Deli dura	High diversity in Nigerian accessions with maximum rare alleles; Deli dura lost 36 more alleles than natural populations.	[58]

(Continued)

Table 3.2 (Continued)

Markers	Germplasm or Somaclonal Variation or Breeding	Conclusion	Reference
40 probes; 20 good enough	54 palms in a breeding program	40 probes covering 60% of genome revealed relationship from the study confirmed the known pedigree.	[59]
4 probes in 6 probe–enzyme combination; PoPgSP12 is highly informative probe.	111 elite breeding palms; 400 clonal lines	Identification of pedigree errors and proper crossing strategy; elimination of individuals with recessive deleterious genes; greater variation from unrelated founder population.	[60]
SSR			
5521 ESTs mined; 145 SSRs found in 136 unique ESTs. Primers designed for 94; primers for 10 EST-SSRs were developed.	76 accessions of oil palm from 7 African countries + Deli dura population	Genetic differentiation (FST) among the populations is 0.2492; high genetic diversity at Nigeria, Congo, and Cameroon; some primers are transferable to coconut and Jessenia bataua.	[61]
19,243 ESTs assembled to give 10,258 unique sequences; 629 ESTs were found to contain 722 SSRs.	Primer pairs were designed for 405 unique EST-SSRs, and 15 of these were used to genotype 105 *E. guineensis* and 30 *E. oleifera* accessions. 14 SSRs were polymorphic.	Deli dura lost 45 alleles more than natural populations. EST-SSR primer sEg00126 with (CGC)7 repeats a high degree of sequence conservation across the palms.	[62]

Table 3.2 (Continued)

Markers	**Germplasm or Somaclonal Variation or Breeding**	**Conclusion**	**Reference**
9 microsatellite markers and 29 quantitative traits	Parental palms (dura and pisifera) and performance of their progenies	Genetic distances based on the microsatellite markers may not be useful for predicting hybrid performance except for mean nut weight and leaf number.	[63]
14 microsatellite loci	318 individuals from 26 accessions from 8 countries	Inbreeding status of germplasm of Deli origin the result of mass selection; positive contribution of the within-population mass selection practiced in the Deli population, and explains the success of Deli × La Mé and Deli × Congo crosses.	[64]
12 informative SSR markers; 11 displayed Mendelian segregation	6 sets of ortets and their respective ramets	Markers identified in the species *Elaeis oleifera* also amplified *Elaeis guineensis* are useful in clonal identification, monitoring line uniformity between and within clones, and detecting culture mix-up.	[65]
21 SSR loci of (GA)*n*, (GT)*n*, and (CCG)*n*	18 E g and 21 Eo accessions and 18 palm speceis	High polymorphism and discrimination capability in Afrcian and American oil palm, and a few are also transferable to a subset of 16 other palm species.	[66]
Isozymes			
7 enzyme loci from 6 enzyme systems	723 accessions of 26 populations from 10 African + Deli dura family	High genetic diversity exists at Nigeria, Cameroon, and Sierra Leone.	[67]

were found to be useful in cultivar identification. High genetic diversity exists in West African germplasm, especially in Nigeria [58,61,67]. Genetic diversity of American oil palm in Caiaue accessions of Amazon Brazil [57] is moderate, whereas it is lowest in accessions from Ghana [58]. Most of the diversity studies use Deli dura as a control to compare with other germplasm accessions. Deli dura lacks many rare alleles and is found to have lost 36 [58] to 45 [62] more alleles than natural populations. Economics of marker-aided breeding over conventional breeding indicated time and money savings from using markers in genome-wide selection [70] (Table 3.3).

3.5 Genes and Promoters of Oil Palm

Important genes and promoters are cloned from oil palm and sequenced to improve the traits of interest. Transcription factors, the master switches regulating the genes of the MADS-box family [72], Knotted gene [73], AP2 domain [75], and Myb-1 [79] of oil palm, are known. Other important genes related to oil yield are accD; ACCase [71], a gene-regulating beta-carotene pathway; beta-carotene hydroxylase [74], a gene involved in early isoprenoid pathway; and ketothiolase [76], and they have been studied in detail in oil palm. Ganoderma wilt, a fungal disease, is a serious problem in oil palm. Chitinases play a crucial role in defense against such fungal diseases. Recently, three chitinases were cloned and sequenced in oil palm [77]. Drought and salinity are major abiotic stress factors reducing productivity in oil palm. A proline transport gene was recently analyzed with potential role in salinity resistance [78]. Specific leaf tissue promoters [80] developed from oil palm have a potential role in the cultivation of transgenic plants in addition to mesocarp-specific [26] and metal-inducible [27,28] ones. Accession numbers of the sequences and details of important genes of oil palm are given in Table 3.4.

3.6 Bioinformatics

Bioinformatics is an emerging field combining information technology and biological sciences. Bioinformatics or *in silico* analysis helps to retrieve, handle, store, and analyze the large-scale sequence data. In oil palm, large-scale annotation of sequences led to the identification of many useful genes. Genomic resources, markers, and linkage map position details of oil palm are available as user-friendly, web-enabled databases [82]. MicroRNA (miRNAs) are short snippets of single-stranded RNA molecules of length 20–22 nucleotides encoded in the genomes. miRNAs negatively regulate the target genes by binding at specific sites and post-translational cleavage. Computational tools are helpful in the prediction and identification of miRNA regions from the genomes. Combined homology and structural analysis in sequences of oil palm expressed in apical meristem, immature, and mature flower tissues were found to contain five potential miRNA-encoding sequences [83].

Table 3.3 SNP Markers and Economics of Marker-Aided Breeding

SNP			
5452 EST sequences mined	*In silico* analysis of SNPs and indels in sequences	1180 SNP sites and 137 indel polymorphism were found; mesocarp had high frequency of 2.91 SNPs and indels per 100 bp.	[68]
24 SNP-based CAPS markers; 6 markers were informative	380,000 oil palm genomic sequences mined and 20,000 good-quality oil palm SNPs identified; restriction enzyme sites were identified.	317 samples from 11 African and 3 South and Central American countries	[69]
Genome-wide selection			Price '03
Economics of MARS (marker-assisted reciprocal recurrent selection)	Cost per unit gain was 26–57% lower with genome-wide selection than with phenotypic selection when markers cost US$1.50 per data point (SSR markers), and 35–65% lower when markers cost $0.15 per data point (SNP, DArT markers).	Genome-wide selection saved time and money in oil palm. With population sizes of $N = 50$ or 70, time per unit gain was 11–23 years with genome-wide selection and 14–25 years with $N = 50$ in oil palm; genome-wide selection is superior to MARS and phenotypic selection in terms of gain per unit cost and time.	[70]

Table 3.4 Genes and Promoters Cloned in Oil Palm

	Gene or Promoter	Process Involved	Accession No.	Conclusion	Reference
1	Biotin carboxylase (accD); acetyl Co-A carboxylase (ACCase)	Lipid accumulation	DQ004687 and AAY86362	Oil palm yield is correlated with expression level of accD gene and biotin carboxylase (accC).	[71]
2	13 MADS-box genes of 5 families SQUAMOSA, GLOBOSA, AGAMOUS-like2, DEFICIENS	Flower formation	AF 411840–AF 411848	Specificity of MADS-box genes in vegetative tissues floral organs groups. ABCDE model of flower structure.	[72]
3	Class I Knotted-like homeobox (KNOX) transcription factors OPKN1 (EgKnox1)	Shoot apical meristem function	DQ890420	Different modes of leaf dissection and leaf morphology.	[73]
4	Beta-carotene hydroxylase (*Chyb*) enzyme of E.o	Carotenoid pathway gene in American oil palm	EU057623	51 fully conserved amino acid residues in *Chyb* from the flowering (monocot and dicot) plants and algal members, *EoChyb* cDNA sequence is 1414 bp in size.	[74]
5	AP2 domain AINTEGUMENTA-like (AIL) and BABYBOOM (BBM)	Transcription factor regulating somatic embryogenesis	AY691196	cDNA of 1786 bp named EgAP2-1 in size	[75]

Table 3.4 (Continued)

	Gene or Promoter	Process Involved	Accession No.	Conclusion	Reference
6	Ketothiolase	Early isoprenoid pathway		415 amino acids 1.5 kb cDNA oil palm mesocarp 8–11 WAA	[76]
7	Serine-threonine protein kinase		DQ267436	Kinase during tissue culture 1.3 kb long 346 amino acids	[33]
8	Promoter region of four embryo-genically up regulated genes	Embryogenesis		Egpk1, EgHOX1, EgPER1, OPSC10	[31]
9	Chitinase	Ganoderma fungal resistance	GU301271	Transcribed gene during fungal infection	[77]
10	Proline/betaine transporter gene and promoter region	Stress resistance to salt	AB597035–AB597036	The transporter gene exhibited high similarity to Bet/ProT genes especially of rice ProT1. Exogenously applied glycinebetaine was taken up rapidly by roots.	[78]

3.7 Haploid Technologies

Inbred lines can capture and enhance the efficiency of heterosis breeding. Development of inbred lines in a perennial crop takes many years and requires experimental space. The use of haploids and doubled haploids as parents for F1 variety production is a potential area of interest in oil palm. Spontaneous occurrence of haploids at very, very low frequencies is reported. Flow cytometric analysis in oil palm provides the quantification of nuclear DNA content. The genome size of African oil palm is $2C = 3.76 \pm 0.09$ pg, which means a genome size of $3.4'10^9$ bp [84]. A high-throughput screen involving morphological traits, molecular (microsatellite) markers, and flow cytometry [84] is developed [85] to identify spontaneously formed haploid (H) and doubled haploid (DH) palms. Nearly 1000 haploids including a single doubled haploid plant were located using the method [85].

3.8 Road Map for Oil Palm Genomics

Genetics, genomics, and transgenic technologies have been exploited to a greater extent in oil palm than in any other cultivated palm. Studies of transcripts in somatic embryogenesis, mantled disorder, mescoarp, and other tissues are exhaustive in oil palm. Molecular markers and genome mapping of shell thickness and mesocarp traits are an important advancement in oil palm. Identification of markers associated with other traits such as dwarf habit and disease resistance is a future priority in oil palm. Such knowledge complements the heterosis of shell thickness exploited through conventional breeding. Haploid technologies coupled with genome-wide selection using markers can bring additional yield enhancement in the highest oil-bearing palm crop. MAP-based cloning of a genomic fragment nearing the shell thickness locus at the terminal region of chromosome 4 is an important future genomics priority. A molecular marker kit or DNA chips for identification of tenera seedlings with 100% precision is the next immediate research priority. Whole genome shotgun sequencing [86] is in progress, which can provide knowledge of the complete genome for oil palm, the next palm species being studied. Research on transcription factors, genes, and promoters cloned from oil palm gives additional tools to oil palm improvement as markers or accessories in transgenics. Studies in retrotransposons [87] and DNA methylation [88] offer additional knowledge of markers in the assessment of diversity in oil palm.

Comprehensive genomic information available in oil palm and a complete genome of date palm offer promise in comparative genomics of other palms. By employing sequence conservation, researchers can transfer the knowledge to other cultivated palms. A few molecular markers have already offered promise in other palm species. In the near future, we anticipate obtaining the complete genome sequence of oil palm and sequence information of many genes and promoters. Genes and promoters of defense-related traits offer protection of oil palm from biotic and abiotic stress factors. Proteomics investigations in important processes of oil palm (ganoderma infected, drought affected) are another priority that is already in progress for date palm. Oil palm genomics offer a model and inspiration for enhancement of yield in other palms.

References

1. Jalani BS, Cheah SC, Rajanaidu N, Darus A. Improvement of palm oil through breeding and biotechnology. *JAOCS*. 1997;**74**:1451–1455.
2. Rival A, Jaligot E. Oil palm biotechnologies are definitely out of infancy. Oléagineux. *Corps Gras, Lipides*. 2010;**17**:368–374.
3. Mayes S, Hafeez F, Price Z, MacDonald D, Billotte N, Roberts J. Molecular research in oil palm, the key oil crop for the future. Moore PH, Ming R, Delmer DP, editors. Genomics of Tropical Crop Plants. Plant Genetics and Genomics: Crops and Models, vol. 1. : Springer Verlag; 2008, p. 371–404.
4. Soh AC, Wong TY, Hor CC, Tan PS, Chew PS. Oil palm In: Oil Crops, Handbook of Plant Breeding, 2010, 4, p. 333–367. DOI: 10.1007/978-0-387-77594-4_11.
5. Sambandamurthi R, Singh R, Kadir APG, Abdullah MO, Kushairi A. Opportunities for oil palm via breeding and biotechnology. In: Jain SM, Priyadarshini PM, eds. *Breeding Plantation Tree Crops: Tropical Tree Species*. Berlin: Springer Science+Business Media LLC; 2009. p. 377–422.
6. Murphy DJ. Oil palm: future prospects for yield and quality improvements. *Lipid Technol*. 2009;**21**:257–260.
7. Parveez GKA, Masri MM, Zainal A, Masri MM, Masani A, Yunus M, et al. Transgenic oil palm: production and projection. *Biochem. Soc. Trans*. 2000;**28**:969–972.
8. Lee M, Yeun L, Abdullah R. Expression of bacillus thuringiensis insecticidal protein gene in transgenic oil palm. *Electron. J. Biotechnol*. 2006;**9**:117–126.
9. Omidvar V, Abdullah SNA, Marziah M, Maheran AA. A transient assay to evaluate the expression of polyhydroxybutyrate genes regulated by oil palm mesocarp-specific promoter. *Plant Cell Rep*. 2008;**27**:1451–1459.
10. Beinaert A, and Vanderweyen R. (1941). Contribution à l'étude génétique et biométrique des variétés d'Elaeis guineensis Jacq. Publications de l'Institut National pour l'étude Agronomique du Congo Belge, série scientifique 27.
11. Hartley CWS. The oil palm, 2nd ed. London: Longman; 1988.
12. Bhasker S, Mohankumar C. Association of lignifying enzymes in shell synthesis of oil palm (*Elaeis guineensis* Jacq.) fruit *dura* variety. *Indian J. Exp. Biol*. 2001;**39**:160–164.
13. Kishor Mohan TC, Pillai RSN, Mohankumar C. Activity of lignin biosynthesis enzymes during fibre ring and shell formation of oil palm (*Elaeis guineensis* Jacq.) fruit–*tenera* variety. *Int. J. Oil Palm*. 2003;**3 and 4**:11–15.
14. Sathish DK, Mohankumar C. RAPD markers for identifying oil palm (*Elaeis guineensis* Jacq.) parental varieties (*dura* & pisifera) and the hybrid *tenera*. *Indian J. Biotechnol*. 2007;6:354–358.
15. Moretzsohn MC, Nunes CDM, Ferreira ME, Grattapaglia D. RAPD linkage mapping of the shell thickness locus in oil palm (*Elaeis guineensis* Jacq). *Theor. Appl. Genet*. 2000;**100**:63–70.
16. Billotte N, Marseillac N, Risterucci AM, Adon B, Brottier P, Baurens FC, et al. Microsatellite-based high-density linkage map in oil palm (*Elaeis guineensis* Jacq.). *Theor. Appl. Genet*. 2005;**110**:754–765.

17. Mayes S, Jack PL, Corley RH, Marshall DF. Construction of a RFLP genetic linkage map for oil palm (*Elaeis guineensis* Jacq.). *Genome*. 1997;**40**:116–122.
18. Thawaro S, Te-chato S. Verification of legitimate tenera oil palm hybrids using SSR and propagation of hybrids by somatic embryogenesis. *Songklanakarin J. Sci. Technol*. 2010;**32**:1–8.
19. Thawaro S, Te-chato. S. Application of molecular markers in the hybrid verification and the assessment of somaclonal variation from *in vitro* propagated oil palm. *Sci. Asia*. 2009;**35**:142–149.
20. Bourgisa F, Kilaruc A, Caod X, Ngando-Ebonguee GF, Driraf N, Ohlrogged JB, et al. (2011). Comparative transcriptome and metabolite analysis of oil palm and date palm mesocarp that differ dramatically in carbon partitioning. Proc. Natl. Acad. Sci., USA. http://www.pnas.org/content/early/2011/06/23/1106502108.full.pdf (in press).
21. Tranbarger TJ, Dussert S, Joet T, Argout X, Summo M, Champion A, et al. (2011). Regulatory mechanisms underlying oil palm mesocarp maturation, ripening and functional specialization in lipid and carotenoid metabolism. *Plant Physiol*. 2011;**156**:564–584.
22. Nurniwalis AW, Suhaimi N, Siti Nor Akmar A, Aminah S, Mohamad Arif MA. Gene discovery via expressed sequence tags from the oil palm mesocarp. *J. Oil Palm Res*. 2008;**2**:87–96. (Spl Issue on Malaysia-MIT Biotechnology Partnership program)
23. San CT, Shah FH. Differential gene expression and characterization of tissue-specific cDNA clones in oil palm using mRNA differential display. *Mol. Biol. Rep*. 2005;**32**:227–235.
24. Xu L, Yuan Y, Zhang L, Wan L, Zheng Y, Zhou P, et al. (2011). Identification and characterization of differential gene expression in the mesocarp and kernel of oil palm nuts using suppression subtractive hybridization. Tree Genet. Genom. 7, 999–1010. DOI: 10.1007/s11295-011-0390-6.
25. Shah FH, Cha TS. A mesocarp-and species-specific cDNA clone from oil palm encodes for sesquiterpene synthase. *Plant Sci*. 2000;**154**:153–160.
26. Siti Nor Akmar A, Zubaidah A. Mesocarp specific metallothionein like gene promoter for genetic engineering of oil palm. *J. Oil Palm Res*. 2008;2:1–8. (Spl Issue on Malaysia-MIT Biotechnology Partnership program)
27. Kamaladini H, Siti Nor Akmar A, Aziz MA. Metal inducible activity of the oil palm metallothionein-like gene promoter (MT3-A) in prokaryotes. *J. Biosci. Bioeng*. 2011;**111**:217–225.
28. Omidvar V, Abdullah SNA, Izadfard A, Ho CL, Mahmood M. The oil palm metallothionein promoter contains a novel AGTTAGG motif conferring its fruit-specific expression and is inducible by abiotic factors. *Planta*. 2010;**232**:925–936.
29. Voelker T. Secrets of palm oil biosynthesis revealed. *Proc. Natl. Acad. Sci., USA*. 2011;**108**:12193–12194.
30. Low ETL, Alias H, Boon SH, Shariff EM, Tan CYA, Ooi LCL, et al. Oil palm (*Elaeis guineensis* Jacq.) tissue culture ESTs: identifying genes associated with callogenesis and embryogenesis. *BMC Plant Biol*. 2008;**8**:62.
31. Ooi SE, Ishak F, Abdullah MO. Cloning and *in silico* analysis of highly expressed genes in oil palm embryogenic cultures. *J. Oil Palm Res*. 2010;**22**:856–868.
32. Roowi SH, Ho CL, Alwee SSRS, Abdullah MO, Napis S. Isolation and characterization of differentially expressed transcripts from the suspension cells of oil palm (*Elaeis guineensis* Jacq.) in response to different concentration of auxins. *Mol. Biotechnol*. 2010;46:1–19.
33. Ooi SE, Harikrishna K, Abdullah MO. Isolation and characterisation of a putative serine-threonine kinase expressed during oil palm tissue culture. *J. Oil Palm Res*. 2008;**1**:14–22. (Spl Issue on Malaysia-MIT Biotechnology Partnership program)

34. Lin HC, Morcillo F, Dussert S, Dubreull CT, Tregear JW, Tranbarger TJ. Transcriptome analysis during somatic embryogenesis of the tropical monocot *Elaeis guineensis* evidence for conserved gene function in early development. *Plant Mol. Biol.* 2009;**70**:173–192.
35. Chan PL, Ma LS, Low ETL, Shariff EM, Ooi LCL, Cheah SC, et al. Normalized embryoid cDNA library of oil palm (*Elaeis guineensis*). *Electron. J. Biotechnol.* 2010;I13.
36. Beule T, Camps C, Debiesse S, Tranchant C, Dussert S, Sabau X, et al. Transcriptome analysis reveals differentially expressed genes associated with the mantled homeotic flowering abnormality in oil palm (*Elaeis guineensis* Jacq.). *Tree Genet. Genomes.* 2010;**7**:169–182.
37. Rival A, Jaligot E, Beule T, Finnegan EJ. Isolation and expression analysis of genes encoding MET, CMT, and DRM methyltransferases in oil palm (*Elaeis guineensis* Jacq.) in relation to the 'mantled' somaclonal variation. *J. Exp. Bot.* 2008;**59**:3271–3281.
38. Ho CL, Kwan YY, Choi MC, Tee SS, Ng WH, Lim KA, et al. Analysis and functional annotation of expressed sequence tags (ESTs) from multiple tissues of oil palm (*Elaeis guineensis* Jacq.). *BMC Genom.* 2007;**8**:381.
39. Singh R, Cheah SC. Differential gene expression during flowering in oil palm. *Plant Cell Rep.* 2000;**19**:804–809.
40. Jouannic S, Argout X, Lechauve F, Fizames C, Borgel A, Morcillo F, et al. Analysis of expressed sequences tags from oil palm (*Elaeis guineensis* Jacq). *FEBS Lett.* 2005;**579**:2709–2714.
41. Bertossi FA, Chabrillange N, Duval Y, Tregar T. Contrasting globulin and cysteine proteinase gene expression patterns reveal fundamental developmental differences between zygotic and somatic embryos of oil palm. *Tree Physiol.* 2011;**28**:1157–1167.
42. Le VT, Sarpan N, Huynh K, Ooi SE, Napis S, Ho CL, et al. A novel transcript of oil palm (*Elaeis guineensis* Jacq.), Eg707, is specifically upregulated in tissues related to totipotency. *Mol. Biotechnol.* 2011;**48**(June (2)):156–164.
43. Jaligot E, Adler S, Debladis E, Beulé T, Richaud F, Ilbert P, et al. (2011). Epigenetic imbalance and the floral developmental abnormality of the in vitro-regenerated oil palm *Elaeis guineensis*. Ann Bot. (2011) doi: 10.1093/aob/mcq266 (in press).
44. Chua KL, Singh R, and Cheah SC. (2001). Construction of oil palm (*Elaeis guineensis* Jacq.) linkage maps using AFLP markers. In *Cutting-edge technologies for sustained competitiveness: Proceedings of the 2001 PIPOC International Palm Oil Congress*, Agriculture Conference by International Palm Oil Congress (2001: Petaling Jaya, Selangor) Kuala Lumpur: Malaysian Palm Oil Board (MPOB) August 20–22, 2001 (pp. 461–467). Kuala Lumpur, Malaysia.
45. Ying ST, Zaman FQ, Ho CL, Maizura I, Rao V. Flanking AFLP markers for the virescence trait in oil palm. *J. Oil Palm Res.* 2007;**19**:381–392.
46. Singh R, Tan SG, Panandam J, Rahman RA, Cheah SC. Identification of cDNA-RFLP markers and their use for molecular mapping in oil palm (*Elaeis guineensis* J.). *Asia Pac. J. Mol. Biol. Biotechnol.* 2008;**16**:53–63.
47. Rohde W, Sniady V, Herrán A, Estioko L, Sinje S, Marseillac N, et al. (2001). Construction and exploitation of high-density DNA marker and physical maps in the perennial tropical oil crops coconut and oil palm: from biotechnology towards marker-assisted breeding. (www.tropentag.de/2002/abstracts/full/44.pdf).
48. Billotte N, Jourjon MF, Marseillac N, Berger A, Flori A, Asmady B, et al. QTL detection by multi-parent linkage mapping in oil palm (*Elaeis guineensis* Jacq.). *Theor. Appl. Genet.* 2010;**120**:1673–1687.
49. Singh R, Tan SG, Panandam J, Rahman RA, Ooi LC, Low ETL, et al. Mapping quantitative trait loci (QTLs) for fatty acid composition in an interspecific cross of oil palm. *BMC Plant Biol.* 2009;**9**:114.

50. Rance KA, Mayes S, Price Z, Jack PL, Corley RHV. Quantitative trait loci for yield components in oil palm (*Elaeis guineensis* Jacq.). *Theor. Appl. Genet.* 2001;103:1302–1310.
51. Kularatne RS, Shah F, Rajanaidu N. Estimation of genetic diversity in some African germplasm collection of oil palm (*Elaeis guineensis* Jacq.) as detected by AFLP markers. *Asia Pac. J. Mol. Biol. Biotechnol.* 2000;**8**:27–36.
52. Purba AR, Noyer JL, Baudouin L, Perrier X, Hamon S, Lagoda PJL. A new aspect of genetic diversity of Indonesian oil palm (*Elaies guinensis* Jacq.) revealed by isozyme and AFLP markers and its consequences for breeding. *Theor. Appl. Genet.* 2000;**101**:956–961.
53. Moretzsohn MC, Ferreira MA, Amaral ZPS, Coelho PJA, Grattapaglia D, Ferreira ME. Genetic diversity of Brazilian oil palm germplasm collected in Amazon forest. *Euphytica.* 2002;**124**:35–45.
54. Shah FH, Rashid O, Simons AJ, Dunsdon A. The utility of RAPD markers for the determination of genetic variation in oil palm (*Elaeis guineensis* Jacq.). *Theor. Appl. Genet.* 1994;**89**:713–718.
55. Junmag S, Nualsri C, and Eksomtramage T. Genetic variation and phylogenetic relationships in oil palm (*Elaeis guineensis* Jacq.) based on RAPD analysis. Songklanakarin J. Sci. Technol. 2005;**27**:473–485.
56. Kanop L, Nugkaew A, Phongdara A, Chotigeat W, Asawatreratankul P, Korawia C, et al. (2005). Identification of polymorphic oil palm (*Elaeis guineensis* Jacq.) using microsatellite, nuclear gene marker and RAPD. 28th Congress on Science & Technology of Thailand, Bangkok Thailand.
57. Rival A, Bertrand L, Beule T, Combes MC, Trouslot P, Lashermes P. Suitability of RAPD analysis for the detection of somaclonal variants in oil palm (*Elaeis guineensis* Jacq.). *Plant Breed.* 1998;**117**:73–76.
58. Maizura I, Rajanaidu N, Zakri A, Cheah S. Assessment of genetic diversity in oil palm (*Elaeis guineensis* Jacq.) using restriction fragment length polymorphism (RFLP). *Genet. Resour. Crop Evol.* 2006;**53**:187–195.
59. Mayes S, Jack PL, Corley RH. The use of molecular markers to investigate the genetic structure of an oil palm breeding programme. *Heredity.* 2000;**85**:288–293.
60. Mayes S, James CM, Horner SF, Jack PL, Corley RHV. The application of restriction fragment length polymorphism for the genetic fingerprinting of oil palm (*E. guineensis* Jacq.). *Mol. Breed.* 1996;**2**:175–180.
61. Singh R, Zaki MN, Ting CN, Rosli R, Tan GS, Low LE, et al. Exploiting an oil palm EST database for the development of gene-derived SSR markers and their exploitation for assessment of genetic diversity. *Biologia.* 2008;**63**:227–235.
62. Ting NC, Noorhariza MZ, Rozana R, Low ET, Ithnin M, Cheah SC, et al. SSR mining in oil palm EST database: application in oil palm germplasm diversity studies. *J. Genet.* 2010;**89**:135–145.
63. Abdullah N, Yusop MR, Ithnin M, Saleh G, Latif MA. Genetic variability of oil palm parental genotypes and performance of its' progenies as revealed by molecular markers and quantitative traits. *C.R. Biologies.* 2011;**334**:290–299.
64. Cochard B, Adon B, Rekima S, Billotte N, de Koutou A, Nouy B, et al. Geographic and genetic structure of African oil palm diversity suggests new approaches to breeding. *Tree Genet. Genomes.* 2009;**5**:493–504.
65. Singh R, Nagappan J, Tan SG, Panandam JM, Cheah SC. Development of simple sequence repeat (SSR) markers for oil palm for oil palm and their application in genetic mapping and fingerprinting of tissue culture clones. *Asia Pac. J. Mol. Biol. Biotechnol.* 2007;**15**(3):121–131.

66. Billotte N, Risterucci AM, Barcelos E, Noyer JL, Amblard P, Baurens FC. Development, characterisation, and across-taxa utility of oil palm (*Elaeis guineensis* Jacq.) microsatellite markers. *Genome*. 2001;**44**:413–425.
67. Hayati A, Wickneswari R, Maizura I, Rajanaidu N. Genetic diversity of oil palm (*Elaeis guineensis* Jacq.) germplasm collections from Africa: implications for improvement and conservation of genetic resources. *Theor. Appl. Genet*. 2004;**108**:1274–1284.
68. Riju A, Chandrasekhar A, Arunachalam V. Mining for single nucleotide polymorphisms and insertions/deletions in expressed sequence tag libraries of oil palm. *Bioinformation*. 2007;**2**:128–131.
69. Li LOC, Wen OP, Maizura I, and Singh R. (2010). Discovery of oil palm snps and their application in the characterisation of oil palm genetic resources and breeding materials. http://www.biotek.gov.my/nbs2010/program/oral/ac/abstract/day3/Leslie%20Ooi%20Cheng-Li%20MPOB.pdf.
70. Wong CK, Bernardo R. Genome wide selection in oil palm: increasing selection gain per unit time and cost with small populations. *Theor. Appl. Genet*. 2008;**116**:815–824.
71. Nakkaew A, Chotigeat W, Eksomtramage T, Phongdara A. Cloning and expression of a plastid-encoded subunit, beta-carboxyltransferase gene (accD) and a nuclear-encoded subunit, biotin carboxylase of acetyl-CoA carboxylase from oil palm (Elaeis guineensis Jacq.). *Plant Sci*. 2008;**175**:497–504.
72. Adam H, Jouannic S, Morcillo F, Richaud F, Duval Y, Tregear JW. MADS box genes in oil palm (*Elaeis guineensis*): patterns in the evolution of the SQUAMOSA, DEFICIENS, GLOBOSA, AGAMOUS, and SEPALLATA subfamilies. *J. Mol. Evol*. 2006;**62**:15–31.
73. Jouannic S, Collin M, Vidal B, Verdeil JL, Tregear JW. A class I KNOX gene from the palm species *Elaeis guineensis* (*Arecaceae*) is associated with meristem function and a distinct mode of leaf dissection. *New Phytol*. 2007;**174**:551–568.
74. Subhash J, Bhore SJ, Kassim A, Loh CY, Shah FH. Analysis of beta-carotene hydroxylase gene cDNA isolated from the American oil-palm (*Elaeis oleifera* HBK) mesocarp tissue cDNA library. *Bioinformation*. 2010;**5**:104–112.
75. Morcillo F, Gallard A, Pillot M, Jouannic S, Aberlenc-Bertossi F, Collin M, et al. EgAP2-1, an AINTEGUMENTA-like (AIL) gene expressed in meristematic and proliferating tissues of embryos in oil palm. *Planta*. 2007;**226**:1353–1362.
76. Teen YJ, Yunus AMM, Parveez AKA, Sambandamurthi R. Activity studies, gene characterization and manipulation of beta ketothiolase of oil palm mesocarp. *J. Oil Palm Res*. 2008;**2**:118–133.
77. Naher L, Ho C-L, Tan SG, Yusuf UK, and Abdullah F. (2011). Cloning of transcripts encoding chitinases from *Elaeis guineensis* Jacq. and their expression profiles in response to fungal infections. Physiol. Mol. Plant Pathol. 10.1016/j.pmpp.2011.06.006.
78. Yamada N, Cha-Um S, Kageyama H, Promden W, Tanaka Y, Kirdmanee C, et al. Isolation and characterization of proline/betaine transporter gene from oil palm. *Tree Physiol*. 2011;**31**:462–468.
79. Teoh WC, Cheah SC, Harikrishna K, Tan SH. Isolation and characterisation of Myb-related genes from oil palm. *J. Plant Biol*. 2003;**46**:95–104.
80. Masani MYA, Parveez GKA, Izawati AMD, Lan CP, SitiNorAkmar A. Construction of PHB and PHBV multiple gene vectors driven by an oil palm leaf specific promoter. *Plasmid*. 2009;**62**:191–200.
81. Willis LB, Lessard PA, Parker JA, O'Brien XM, Sinskey AJ. Functional annotation of oil palm genes using an automated bioinformatics approach. *J. Oil Palm Res*. 2008;**1**:35–43. (Spl Issue on Malaysia-MIT Biotechnology Partnership program)

82. Ruiz M, Rouard M, Raboin LM, Lartaud M, Lagoda P, Courtois B. TropGENE-DB, a multi-tropical crop information system. *Nucleic Acids Res.* 2004;**32**:D364–D367.
83. Nasaruddin N, Harikrishna K, Othman R, Hoon L, Harikrishna J. Computational prediction of microRNAs from oil palm (*Elaeis guineensis* Jacq.) expressed sequence tags. *Asia Pac. J. Mol. Biol. Biotechnol.* 2007;**15**:107–113.
84. Rival A, Beule T, Barre P, Hamon S, Duval Y, Noirot M. Comparative flow cytometric estimation of nuclear DNA content in oil palm (*Elaeis guineensis* Jacq.) tissue cultures and seed-derived plants. *Plant Cell Rep.* 1997;**16**:884–887.
85. Dunwell JM, Wilkinson MJ, Nelson S, Wening S, Sitorus AC, Mienanti D, et al. Production of haploids and doubled haploids in oil palm. *BMC Plant Biol.* 2010;**10**:218.
86. Zieler H, Richardson T, Schwartz A, et al. (2010). Whole-genome shotgun sequencing of the oil palm and Jatropha genomes. In: *Plant & Animal Genomes XVIII Conference*, January 9–13. San Diego, CA.
87. Price Z, Dumortier F, MacDonald W, Mayes S. Characterisation of copia-like retrotransposons in oil palm (*Elaeis guineensis* Jacq.). *Theor. Appl. Genet.* 2002;**104**:860–867.
88. Kubis SE, Castilho AMMF, Vershinin AV, Seymour J, Harrison H. Retroelements, transposons and methylation status in the genome of oil palm (*Elaeis guineensis*) and the relationship to somaclonal variation. *Plant Mol. Biol.* 2003;**52**:69–79.

4 Date Palm

4.1 Introduction

Date palm is grown in an area of 1.2 million hectares mainly in North Africa and the Middle East, with the production of 7 million metric tons of fruits. Egypt, Iran, Saudi Arabia, and the United Arab Emirate (UAE) are the major producing countries, contributing to 57% of the global production (FAOSTAT, 2008). Genomics research complements classical breeding in characterizing the diversity, developing resistance to Bayoud disease, identifying sex of the seedling, using probes for detecting brittle leaf disease, studying proteins involved in major biochemical process, and so on.

4.2 Genetic Diversity using Molecular Markers

4.2.1 Amplified Fragment Length Polymorphism

Amplified fragment length polymorphism (AFLP) is a selective PCR amplification of restriction fragments of the total digest of genomic DNA [1]. EcorI (E) and MseI (M) restriction enzymes with triplet nucleotide adapter sequences are used. AFLP markers generate a large number of scorable bands and were used in the identification of Californian date cultivars [2]. Thirty-one AFLP primers of E-ACC and M-CAC combinations gave bands, and 15 of them were polymorphic [3] in segregating progenies for fruit quality and tolerance to salinity and high humidity. There are many homonyms and synonyms for names of date varieties, owing to the long history of cultivation and germplasm exchanges in North Africa. But AFLP profiles [4] of similarly named varieties (homonyms) of Bint Aisha (Bent Aisha 1) showed large AFLP polymorphisms compared to the other three Bint Aisha accessions [3]. Ammry1 and Ammry2 were not the same accessions, although they have similar names [3]. Hence, AFLP markers [1] are also helpful in predicting origin in mislabeled varieties among the homonyms [4] and identification of [2] genetically similar and dissimilar date palm varieties.

4.2.2 RAPD, SSR, and ISSR Markers

RAPD and ISSR markers were employed in date palm for characterizing the diversity native to different geographical locations and also to discriminate male and female palms, and between specific cultivars and wild (*P. canariensis*) and cultivated (*P. dactylifera*) species. The *Phoenix* genus has many species, of which *P. canariensis* is endemic to the Canary Islands of Spain. *P. dactylifera* palms were also introduced in the same island and are easily crossed with the *P. canariensis*. Hence,

Genomics of Cultivated Palms. DOI: 10.1016/B978-0-12-387736-9.00004-2

seedlings of both the species and their intermediate types are present in the islands in abundance. Operon RAPD primers, especially OPM08 and OPK14, which have species-specific bands, are helpful to discriminate the seedlings [5].

RAPD markers were used to characterize the diversity of germplasm of Egypt [6], Jordan [7], Saudi Arabia [8], and Morocco [9]. High genetic diversity in Tunisian germplasm was demonstrated by the use of RAMCO (a combined marker system of RAPD and ISSR primers) and a set of microsatellite loci [10]. A microsatellite kit for characterizing the date palm diversity was developed using 16 loci [11]. These loci are conserved and cross-transferable to 14 species of *Phoenix* and 6 other genera of palms [11,12]. They were used with minor modifications to test lack of gene flow in *P. atlantica* [13] and diversity in date palm germplasm native to Algeria [12], Tunisia [10], Sudan [14], and Qatar [15]. Genetic diversity of date palm is low in Morocco [9], Saudi Arabia [8], and Algeria [12] but rather high in Tunisia [10] mainly due to the history of germplasm or cultivar exchange in the respective countries. Additional details are provided in Table 4.1 on the use of markers at species and in Table 4.2 on date palm diversity at different geographical locations. A comprehensive study of germplasm accession of all the date-growing countries, using a universal set of RAPD, AFLP, ISSR, and SSR loci, should be undertaken to get a consensus of the location of biodiversity hot spots.

4.3 Molecular Biology of Sex Determination

Date palm is a dioecious plant with individual male and female trees. The tree takes 5–7 years to flower, and hence long periods of time are required before the sex of the plant is known. It is easy to propagate the female trees by offshoots or by

Table 4.1 Molecular Markers for Use in Related Species of Phoenix

Species	Marker Systems	Conclusion	Reference
P. canariensis, *P. dactlyifera*, and intermediate types	54 operon RAPDs	OPM08 and OPK14 gave species-specific bands	[5]
11 species of *Phoenix* and 4 palm genera	16 nuclear SSR loci of GA repeat	Cross-transferable to 11 other *Phoenix* species, and a few were transferable to *Elaeis guineensis*, 11 species of *Pritchardia* and *Pritchardiopsis jeanneneyi*, and 6 species of *Astrocaryum*	[11]
P. dactylifera, *P. canariensis*, *P. sylvestris*, *P. atlantica*	15 nuclear SSR loci, 1 plastid minisatellite	*P. atlantica* distinct and isolated for a long time, with no gene flow	[13]
14 species of *Phoenix*	17 SSR loci	Also suitable for 14 other species of *Phoenix*	[12]

Table 4.2 Molecular Markers for Genetic Diversity of Date Palm

Country	Samples Used	Marker Systems	Conclusions	Reference
Egypt	57 samples of 21 known accessions, 9 unknown pedigrees	AFLP	Clarity in mislabeled homonyms and synonyms of cultivars	[4]
Egypt	4 female and 4 male palms	5 RAPD primers	Dendrograms and relationships of cultivars	[6]
Jordan	5 varieties	Operon series A to Z RAPDs	Cultivar specific bands; OPZ07, OPZ10 polymorphic	[7]
Saudi Arabia	13 cultivars	Operon series A to G RAPDs	37 polymorphic primers; narrow genetic diversity; Barney cultivar distinct	[8]
Morocco	37 accessions of Morocco; 6 from Iraq and Tunisia	123 Operon RAPDs	19 were polymorphic; low diversity in Morocco germplasm	[9]
Tunisia	30 cultivars; 10 males	18 RAMPOs (3 RAPD × 6 ISSRs)	Deglet Nour, Ammari, and Khessab formed a group; continuous genetic variation; clustering irrespective of geographic origin or sex	[16]
Tunisia	49 cultivars	16 pdCir SSR loci	High genetic diversity in Tunisian germplasm; no amplification in locus mPdCIR48; locus mPdCIR90 displays 4 unique alleles and 7 alleles in subgroup Tozeur oasis	[10]
Sudan	37 male and 23 female accessions; and 8 female from Morocco	16 pdCir SSR loci	Loci mpdCIR044 is selectively neutral to dry date group; high genetic diversity in Sudan germplasm; weak clustering, no specific grouping	[14]
Qatar	15 cultivars	16 pdCir SSR loci	6 primers—no amplification; 10 primers—single band; Abu Main distinct cultivar; Barhee and Sultane are closely related	[15]
Algeria	31 accessions of Algeria and USDA California	17 SSR loci	High in-breeding in the accessions of Algeria	[12]
Mauritius	10 cultivars	Mitochondrial plasmids S and R	R band 1345 bp; S band 1454 bp; intracultivar variability in Ahmar; strict maternal inheritance of mitochondria	[17]

Table 4.3 Sex-Specific Markers in Date Palm

Materials	Methodology	Conclusions	Reference
Root meristem	Chromomycin staining; confirmation by confocal microscopy flow cytometry	Occurrence of sexual chromosomes carrying distinctive nucleolar heterochromatin	[19]
4 dry cultivars, 3 superior pollinators	RAPD and ISSR	Male- or female-specific bands in OPA10, OPA12, OPD10 5 female-specific ISSR bands	[20]
5 females, 3 males	De novo genome sequencing and SNP genotyping	1605 SNPs that segregated with gender	[21]

using tissue culture techniques. But in breeding experiments, predicting the sex of the seedling well in advance is important, to save the time, land, and labor required to maintain the desired sex of the plant. In date palm, developmental arrest of sterile sex organs occurs by the end of cell division. It is followed by the precocious cell differentiation and development of unisexual flowers [18]. Simple root meristem staining with chromomycin is sufficient to identify the sex of the date palm plant [19]. Three RAPD and five ISSR primers (Table 4.3) were useful in identifying the sex of a date palm plant. These female-specific (OPA10-490, OPA12-750, OPD10-800) and male-specific (OPA12-370, OPD10-675, HB10-1010, HB9-340, HB12-375, 814-590, 844A-920) bands [20] could be cloned and sequenced. These are converted to sequence characterized amplified region (SCAR) markers for further validation and used to identify sex at the seedling stage. Recently, *de novo* genome sequencing of date palm was attempted [21] for five females and three males. Many single nucleotide polymorphisms related to sex were identified from the study.

4.4 Resistance to Bayoud Disease

Bayoud disease of date palm is caused by the fungal pathogen *Fusarium oxysporum* f. sp. *albedinis* (Foa), also known as Fusariosis. The disease first appeared in the Draa Valley [22] in the south of Morocco in 1870. The pathogen enters the plant through the roots, producing foliar withering and leading to the death of the tree. It is the most important disease of the date palm crop and has destroyed nearly 10 million trees in Morocco (two-thirds of plantations) and more than 3 million trees in Algeria; it is also found in Argentina and is a serious problem in other date-producing countries. Two minicircular plasmid-like DNA of size 1346 bp and (U) 1160 bp, sharing homology, were discovered from the mitochondria of a Moroccan variety of date palm [23]. Later, an S plasmid was found in some varieties of 1.45 kbp size [24]. R plasmid evolved from S plasmid by a recombination event of 26 bp imperfect short direct repeats. These two plasmids differed by a size of a 109 bp segment and were amplified by [25] using Oli1 and Oil2 primers. Six Moroccan varieties resistant to the disease had only R plasmid and lacked S plasmid. These varieties had poor fruit quality. Hence, variation in the mitochondrial plasmid was found to be a potential candidate marker for resistance breeding. However, later the association between the marker and resistance [26] was verified and found only in 77% (seven out of nine) of plants. The remaining two susceptible varieties had the R plasmid or both S and R plasmids [26]. Various biochemical mechanisms, toxins, and the molecular biology of the Bayoud disease resistance of date palm were reviewed recently [27] and provide additional details on the subject.

4.5 Probes for Brittle Leaf Disease

"Brittle leaf disease," or the "Maladie des feuilles cassantes" (MFC), is a disorder affecting date palms in Tunisia. Leaflets of young fronds of the affected palms become brittle and break when flexed and squeezed. Later, the other fronds also develop similar symptoms and the tree dies. The disorder is prevalent in the Nafta oasis of Tunisia [28], Eastern Algeria, and Libya, and the leaves of the affected palms are deficient in Mn [29]. Date palms with brittle leaf disease are associated with the accumulation of two populations of small, chloroplast-encoded RNAs corresponding to two different regions of the date palm chloroplast genome. The regions are rrn5s/trnR and ATPE, coding for 5S rRNA, tRNA–Arg, and *e* subunit of ATP synthase, respectively. Dot blot hybridization of these two RNA molecules gave a positive hybridization signal in samples of affected tissues of any cultivar, with varying severity of symptoms, sampled from different geographical locations. But the signal was not seen in samples from the trees affected by other biotic and abiotic stresses. Hence, the dot blot assay is simple and specific for the disorder and useful for early diagnostic purposes [30]. Isolation and molecular characterization of these two RNAs were analyzed for the properties of circular RNA viroids but were found to be double-stranded RNAs of host origin [31]. Although the exact cause of the disease is still unknown, a rapid detection technique is available that uses the dot blot technique, using an RNA probe specific to date palm.

4.6 Somaclonal Variations, DNA Methylation, and Other Markers

Tissue culturing of date palm is done commercially to produce elite genotypes quickly. It is helpful to overcome the few offshoots by a mother plant over a year. Somaclonal variations do come up, which need to be detected to maintain genetic stability. Tissue culture-derived date palm offshoots, analyzed using 37 RAPD primers and 13 AFLP primer combinations, pointed to less than 3% genetic variation, confirming the genetic stability of tissue culture-derived dry date palm cultivars [32]. Differences in the pattern of methylation of DNA were noticed in the dwarf type of Medjool and the normal tall Medjool variety. Around 2000 genes differed in expression pattern between normal and dwarf [33]. Genes involved were the biosynthesis of gibberellins, auxins, brassinosteroids, and expansins, and WRKY transcription factors were mainly shown in differential expression profiles. About 21 of these genes [33] were verified by quantitative reverse transcriptase PCR (qRT-PCR).

Methylation of cytosine is an important phenomena affecting gene expression in locations where C is methylated. A methylation-sensitive amplified polymorphism (MSAP) technique was developed for date palm, using this feature to analyze the DNA variation in offshoots and mother palm. Hypomethylation of DNA was found in offshoots of date palm compared with mother plants. It may be related to gene expression during offshoot development [34].

DNA amplification fingerprinting (DAF) is another set of markers initially used in Tunisian germplasm [35]. Restricted fragment length polymorphism (RFLP) analysis [36] is used to complement the RAPD markers in cultivar identification of date palm. RFLP using a cDNA probe of a zinc finger protein motif is useful to screen the polymorphism among cultivars and to tag important traits (yield, fruit quality, and resistance to disease) in date palm in different countries where it is grown [37]. Apart from these, molecular markers were helpful to analyze the seedling of a rare ancient date seed excavated [38] from Masada and radiocarbon-dated to the first century C.E. The study provides insights into the historic date culture of the Dead Sea region as representative of an extinct date palm population [38].

4.7 Complete Genomes, SNPs, and Genomic Libraries

The genomic library was constructed from total cellular DNA, previously amplified according to an RAPD [39] procedure and cloning. The resultant recombinant DNA [40] is characterized by a size ranging from 200 to 1600 bp inserts. It constitutes a large number of anonymous probes of potential application in Southern hybridization experiments. It also generates markers for molecular characterization of date palm varieties and those associated with fruit quality, resistance to Bayoud disease, and sex of trees. The complete genome of the chloroplast (cp) of date palm is A + T rich, of 158 kb size, and is sequenced and available now [41]. About 78 SNPs were discovered as major intravarietal polymorphisms within the population of a specific cp genome, most of which were located in genes with vital functions. Based on RNA-sequencing data, 18 polycistronic transcription units and three highly expression-biased genes—atpF, trnA-UGC, and rrn23—along with their tissue specific patterns, were found. The complete cp genome of date palm was compared with the 81 available coding sequences of the oil palm cp genome to locate the intraSNPs. Nevertheless, date palm and oil palm belong to two different genera and 19 sites, whose cp minor genotypes in date palm became major genotypes in oil palm, or vice versa. These SNPs have potential application in markers for palms.

A draft nuclear genome for a Khalas variety female date palm [21] was assembled of ~380Mb size, covering 60% of the genome and 90% of the genes. By sequencing eight other cultivars, millions of SNPs were identified; some of them are also useful in discriminating cultivars and to identify sex [21]. Draft date palm was mined for simple sequence repeats and was revealed to contain >32,000 SSRs [42]. Genomic regions containing Class I SSRs (repeat size >20) were annotated, of which many were either only hypothetical proteins or transcription factors, protein kinases zinc finger proteins [42], and so on. The draft nuclear and plastid genome of date palm is a milestone in palm genomics and the first of its kind in the Arecales. It also offers scope for further research on comparative genomics and computational genomics of other cultivated palms.

4.8 Proteomics

Proteomics is the study of the sum total of proteins of an organism. Three reports are available on proteomes of date palm, providing insights to its genetic organization [43–45]. Plant defense or stress-associated proteins were induced, while the energy metabolism or photosynthesis-associated proteins were affected during the colonization of endophytic entomopathogenic fungi in date palms grown in field [43]. A myosin heavy chain-like protein was accumulated in date palm grown *in vitro* when inoculated with the same fungi [43]. About 21 proteins were identified during development and germination of zygotic date embryos [44]. Enzymes of glycolysis, carbohydrate synthesis, protein translation, storage proteins, and stress-related proteins were found during development and germination of zygotic date embryos. Storage proteins such as glutelin, similar to that of oil palm, aldol reductase, and glyoxylase I, were some of the important proteins identified during the study [44]. Eight of the 23 identified proteins during somatic and zygotic embryogenesis were part of the glycolysis pathway. Others were of carbohydrate biosynthesis, citrate cycle, ATP synthesis, and amino acid metabolism. Most of the somatic embryo-specific proteins identified belonged to glycolysis pathways, whereas those of the zygotic embryo belonged to storage and stress-related proteins. Four are from stress-related and three are from storage proteins, and three were not assigned any function [45].

An understanding of such basic processes of proteomics sheds light on the importance of storage or stress proteins or those involved in biochemical pathways. However, investigation is needed in important processes such as the differences between poor- and good-quality fruit, tissues infected by Bayoud pathogen in resistant and susceptible cultivars, tissues affected by brittle leaf disorder, and in other areas. Research about the proteins involved in such processes is essential to enhance the knowledge and develop appropriate markers or kits for application in date palm improvement.

4.9 Road Map for Date Palm Genomics

Date palm, *Phoenix dactylifera* L., is a perennial dioecious in North Africa and the Middle East. It is also useful as a component of arid ecosystems and oases, and is used as fruit, fuel, and fiber. Operon RAPD primers, especially OPM08 and OPK14, having species-specific bands and female- or male-specific bands [20], should be cloned and sequenced. Conversion of these specific dominant markers to reliable sequence characterized amplified region (SCAR) markers should be done for further validation and to identify sex or species at the seedling stage. Date palm proteomics is available on endophytic fungal symbiosis, somatic and zygotic embryogenesis, seed development, and germination. Such investigations are urgently needed in important processes such as differentiating between poor- and good-quality fruit, tissues infected by Bayoud pathogen in resistant and susceptible cultivars, tissues affected by brittle leaf disorder, and in other areas. An understanding of the proteins involved in such processes is essential to enhance the knowledge and develop appropriate markers or diagnostic kits for application in date palm improvement. Gaps in research in date genomics include the need for markers for fruit quality, the requirement of a universal set of markers for diversity of global date germplasm, studies on linkage mapping and QTLs for important traits, research on gene cloning governing important traits, and bioinformatics of complete genomes.

References

1. Vos P, Hogers R, Bleeker M, Reijans M, van de Lee T, Hornes M, et al. AFLP: a new technique for DNA fingerprinting. *Nucl. Acid Res.* 1995;**23**:4407–4414.
2. Cao BR, Chao CT. Identification of date (*Phoenix dactylifera*) cultivars in California using AFLP markers. *HortScience*. 2002;**37**:966–968.
3. El-Kharbotly A, El-Mardi M, Al-Saadi N, Al-Mahruki Y. Towards the construction of a genetic map of date palms using the amplified fragment length polymorphism technique (AFLP). *Acta Hort.* 1998:194–207.
4. El-Assar AM, Krueger RR, Devanand PS, Chao CT. Genetic analysis of Egyptian date (*Phoenix dactylifera* L.) accessions using AFLP markers. *Genet. Resour. Crop Evol.* 2005;**52**:601–607.
5. Perez MAG, Castells JC, Sosa PA. Molecular evidence of hybridisation between the endemic *Phoenix canariensis* and the widespread *P. dactylifera* with Random Amplified Polymorphic DNA (RAPD) markers. *Plant Syst. Evol.* 2004;**247**:165–175.
6. Soliman SS, Ali BA, Ahmed MMM. Genetic comparisons of Egyptian date palm cultivars (*Phoenix dactylifera* L.) by RAPD-PCR. *Afr. J. Biotech.* 2003;**2**:86–87.
7. Rawashdeh I, Amri A. Genetic characterization of date palm varieties using RAPD markers. *Jordan J. Agric. Sci.* 2006;**2**:234–241.
8. Al-Khalifah NS, Askari E. Molecular phylogeny of date palm (*Phoenix dactylifera* L.) cultivars from Saudi Arabia by DNA fingerprinting. *Theor. Appl. Genet.* 2003;**107**:1266–1270.
9. Sedra MH, Lashermes P, Trouslot P, Combes MC, Hamon S. Identification and genetic diversity analysis of date palm (*Phoenix dactylifera* L.) varieties from Morocco using RAPD markers. *Euphytica*. 2004;**103**:75–82.
10. Zehdi S, Trifi M, Billotte N, Marrakchi M, Pintaud JC. Genetic diversity of Tunisian date palms (*Phoenix dactylifera* L.) revealed by nuclear microsatellite polymorphism. *Hereditas*. 2004;**141**:278–287.

11. Billotte N, Marseillac N, Brottier P, Noyer JL, Jacquemoud-Collet JP, Moreau C, et al. Nuclear microsatellite markers for the date palm (*Phoenix dactylifera* L.): characterization and utility across the genus *Phoenix* and in other palm genera. *Mol. Ecol. Notes.* 2004;**4**:256–258.
12. Akkak A, Scariot V, Marinoni T, Boccacci P, Beltramo C, Botta R. Development and evaluation of microsatellite markers in *Phoenix dactylifera* L. and their transferability to other *Phoenix* species. *Biol. Plant.* 2009;**53**:164–166.
13. Henderson SA, Billotte N, Pintaud J-C. Genetic isolation of Cape Verde Island *Phoenix atlantica* (Arecaceae) revealed by microsatellite markers. *Cons. Genet.* 2006;**7**:213–223.
14. Elshibli S, Korpelainen H. Microsatellite markers reveal high genetic diversity in date palm (*Phoenix dactylifera* L.) germplasm from Sudan. *Genetica.* 2008;**134**:251–260.
15. Ahmed TA, Al-Qaradawi AY. Molecular phylogeny of Qatari date palm genotypes using simple sequence repeats markers. *Biotechnology.* 2009;**8**:126–131.
16. Rhouma S, Dakhlaoui-Dkhil S, Salem AOM, Zehdi-Azouzi S, Rhouma A, Marrakchi M, et al. Genetic diversity and phylogenic relationships in date-palms (*Phoenix dactylifera* L.) as assessed by random amplified microsatellite polymorphism markers (RAMPOs). *Sci. Hort.* 2008;**117**:53–57.
17. Salem AOM, Rhouma A, Zehdi S, Marrakchi M, Trifi M. Molecular characterization of Mauritanian date palm cultivars. *Biol. Plant.* 2007;**51**:169–172.
18. Daher A, Adam H, Chabrillange N, Collin M, Mohamed N, Tregear JW, et al. Cell cycle arrest characterizes the transition from a bisexual floral bud to a unisexual flower in *Phoenix dactylifera* L. *Ann. Bot.* 2010;**106**:255–266.
19. Yakovlev SS, Benmalek S, Cerbah M, Coba de la Pefia T, Bounaga N, Brown SC, et al. Chromosomal sex determination and heterochromatin structure in date palm. *Sex Plant Reprod.* 1996;**9**:127–132.
20. Younis RAA, Ismail OM, Soliman SS. Identification of sex-specific DNA markers for date palm (*Phoenix dactylifera* L.) using RAPD and ISSR techniques. *Res. J. Agric. Biol. Sci.* 2008;**4**:278–284.
21. Al-Dous EK, George B, Al-Mahmoud ME, Al-Jaber MY, Wang H, Salameh YM, et al. De novo genome sequencing and comparative genomics of date palm (*Phoenix dactylifera*). *Nat. Biotech.* 2011;**29**:521–527.
22. Foex E, Vanssiere P. Les maladies du dattier au Maroc. *J. Agric. Trop.* 1919;**162**: 336–339.
23. Benslimane AA, Rode F, Quétier F, Hartmann C. Characterization of two minicircular plasmid-like DNAs isolated from date-palm mitochondria. *Curr. Genet.* 1994;**26**:535–541.
24. Benslimane AA, Hartmann C, Ouenzar B, Rode A. Intracellular recombination of a mitochondrial minicircular plasmid-like DNA of date palm mediated by a set of short direct repeat sequences. *Curr. Genet.* 1996;**29**:591–593.
25. Ouenzar B, Trifi M, Bouachrine B, Hartmann C, Marrakchi M, Benslimane AA, et al. A mitochondrial molecular marker of resistance to Bayoud disease in date palm. *Theor. Appl. Genet.* 2001;**103**:366–370.
26. Trifi M, Rhouma A, Marrakchi M. (200!). A molecular marker of date-palm (*Phoenix dactylifera* L.) resistance to Bayoud disease. Proc. second International Conference on Date Palms (Al-Ain, UAE, March 25–27, 2001), United Arab Emirate University UAE. pages 604–612. http://www.pubhort.org/datepalm/datepalm2/datepalm2_75.pdf

27. El Modafar C. Mechanisms of date palm resistance to Bayoud disease: current state of knowledge and research prospects. *Physiol. Mol. Plant Pathol.* 2010;**74**:287–294, 10.1016/j.pmpp.2010.06.008.
28. Triki MA, Zouba A, Khoualdia O, Ben Mahamoud O, Takrouni ML, Garnier M, et al. "Maladie des feuilles cassantes" or brittle leaf disease of date palms in Tunisia: biotic or abiotic disease?. *J. Plant Pathol.* 2003;**85**:71–79.
29. Namsi A, Montarone M, Serra P, Ben Mahamoud O, Takrouni ML, Zouba A, et al. Manganese and brittle leaf disease of date palm trees. *J. Plant Pathol.* 2007;**89**:49–60.
30. Namsi A, Marqués J, Fadda Z, Takrouni ML, Ben Mahamoud O, Zouba A, et al. Diagnosis of "maladie des feuilles cassantes" or brittle leaf disease of date palms by detection of associated chloroplast encoded double stranded RNAs. *Mol. Cell. Probe.* 2006;**20**:366–370.
31. Marqués J, Fadda ZGN, Duran-Vila N, Flores R, Bové JM, Daròs JA. A set of novel RNAs transcribed from the chloroplast genome accumulates in date palm leaflets affected by brittle leaf disease. *Phytopathology*. 2008;**98**:337–344.
32. Saker MM, Adawy SS, Mohamed AA, El-Itriby HA. Monitoring of cultivar identity in tissue culture-derived date palms using RAPD and AFLP analysis. *Biol. Plant.* 2006;**50**:198–204.
33. Ramanovsky A. (2010). Molecular analysis of somaclonal variants in date palm (*Phoenix dactylifera* L.). Master's thesis submitted to Bar-Ilan University. Ramat Gan, Israel, 80 pp.
34. Fang JG, Chao CT. Methylation-sensitive amplification polymorphism in date palms (*Phoenix dactylifera* l.) and their off-shoots. *Plant Biol. (Stuttg.)*. 2007;**9**:526–533.
35. Trifi M, Rhouma A, Marrakchi M. Phylogenetic relationships in Tunisian date palm (*Phoenix dactylifera* L.) germplasm collection using DNA amplification fingerprinting. *Agronomie*. 2000;**20**:665–671.
36. Cornicquel B, Mercier L. Date-palm (*Phoenix dactylifera* L.) cultivar identification by RFLP and RAPD. *Plant Sci.* 1994;**101**:163–172.
37. Cornicquel B, Mercier L. Identification of date palm (*Phoenix dactylifera* L.) cultivars by RFLP: partial characterization of a cDNA probe that contains a sequence encoding a zinc finger motif. *Int. J. Plant Sci.* 1997;**158**:152–156.
38. Sallon S, Solowey E, Cohen Y, Korchinsky R, Egli M, Woodhatch I, et al. Germination, genetics, and growth of an ancient date seed. *Science*. 2008;**320**:1464.
39. Williams JGK, Hanafey MK, Rafalski JA, Tingey SV. Genetic analysis using random amplified polymorphic DNA markers. *Methods Enzymol.* 1993;**218**:704–740.
40. Sakka H, Trifi M, Ali OMS, Rhouma A, Marrakchim M. Rapid construction of a random genomic library from date-palm (*Phoenix dactylifera L.*). *Plant Mol. Biol. Rep.* 2000;**17**:1–7.
41. Yang M, Zhang X, Liu G, Yin Y, Chen K, Yun Q, et al. The complete chloroplast genome sequence of date palm (*Phoenix dactylifera* L.). *PLoS ONE*. 2010;**5**(9):e12762.
42. Manju KP, Arunachalam V. (2010). Genome wide analysis of simple sequence repeats in date palm. *Poster at the international symposium on computational biology, structural bioinformatics and systems biology (Bioinformatica Indica)*. 11–13 January 2010 at Centre for Bioinformatics, University of Kerala, Trivandrum, Kerala, India.
43. Gómez-Vidal S, Salinas J, Tena M, Lopez-Llorca LV. Proteomic analysis of date palm (*Phoenix dactylifera* L.) responses to endophytic colonization by entomopathogenic fungi. *Electrophoresis*. 2009;**30**:2996–3005.

44. Sghaier B, Valledor L, Drira N, Jorrin-Novo J. Proteomics, analysis of the development and germination of date palm (*Phoenix dactylifera* L.) zygotic embryos. *Proteomics.* 2009;**9**:2543–2554.
45. Sghaier B, Drira N, Jorrin-Novo J. Comparative 2-DE proteomic analysis of date palm (*Phoenix dactylifera* L.) somatic and zygotic embryos. *J. Proteomics.* 2009;**73**:161–177.

5 Rattans and Palmyra

5.1 Introduction

Rattans are spiny climbing palms used in making cane furniture and mats, an industry that is growing at US$4 billion trade per year. Rattans belong to the subfamily Calamoideae of the family Arecaceae. The Calamoideae subfamily consists of 22 genera and nearly 650 species [1]. Genomics were useful in phylogeny reconstruction of Calamodeae, and the molecular taxonomy of *Calamus* and related genera is an important research focus. Rattans are harvested from forests that require conservation efforts. Assessment of genetic diversity in rattans in diversity hotspots aids in identifying gene erosion, and priority locations and species for conservation. Sex identification of the *Calamus* seedling is a priority because the plant takes 5 years to enter the reproductive phase. Molecular markers associated with sex in rattan are needed to identify the sex at an early stage.

Palmyra (*Borasssus flabellifer* L.) is a long-lived perennial dioecious palm common in dry tropics, especially in Asia and Africa. Inflorescence sap is tapped from the palms and is used in making sugar and alcoholic beverages. The fruit of the palm is also edible. Molecular markers are useful in analyzing the diversity, trait–marker association in sex, dwarf habit, and high sap yield.

5.2 Molecular Markers for Rattans

5.2.1 Molecular Phylogeny

Members of the Calamoideae subfamily were analyzed with a sample of DNA from 40 or 41 species. Sequence variations in one internal transcribed spacer (ITS) region in a nuclear gene 18S–26S nrDNA [2] cistron and an intron of the ribosomal protein *rps*16 of chloroplast, each of ~900 bp size, were analyzed in these species [2]. ITS regions of the calamoid palms (671 bp size) displayed polymorphism. Using three internal primer sets in the ITS regions, two internal regions, ITS1 and ITS2, of 163–243 bp size, were sequenced in the palms. Mean size of the *rps*16 intron of palms was found to be 870 bp. Calameae was demonstrated as a paraphyletic group including Lepidocaryeae, using these two loci. *Raphia*, the African genus, is found to be a sister clade to the Lepidocaryeae. African and Asian rattans formed two separate monophyletic groups except *Eugeissona* [3]. Success of the nuclear 5S nontranscribed spacer region in classification of the palm genus *Phoenix* [4] inspired the use of this region in DNA for classification and relationship analysis of *Calamus* and five related genera [5]. The amplified PCR product of PIII and PIV primers of the 5S spacer in the rattans ranged from 85 to 413 bp. Results of the 5S nrDNA data found the genus *Calamus* to be a paraphyletic group with four major lineages [5].

Genomics of Cultivated Palms. DOI: 10.1016/B978-0-12-387736-9.00005-4

5.2.2 Molecular Markers for Diversity in Rattans

Six published coconut microsatellite markers [6] were tested for cross-amplification in rattans [7] of four genera, *Korthalsia*, *Zalacca*, *Daemonorops*, and *Calamus*. Two of these markers had shown amplification (Table 5.1). RAPD markers were used in

Table 5.1 Molecular Markers in Rattans

Purpose	Materials and Methods	Conclusion	Reference
Known SSR markers in palms for rattans	CNZ05, CNZ19, CNZ34, CNZ46, CN2A5, and CN1G4 [6]	2 microsatellite primer pairs, CNZ34 and CN2A5, for molecular differentiation of rattans. Lack of polymorphism for CNZ34 and CN2A5 within *C. thwaitesii*.	[7]
RAPD marker-aided diversity in *C. metzianus* and *C. rivalis* in India and Sri Lanka	65 operon RAPD primers (OPA, OPB, OPE, OPAU, OPAW series)	10 RAPD primers (OPAU02, OPA04, OPA18, OPAW07, OPAW09, OPAW10, OPAW20, OPE02, OPE18, and OPB15) good for *Calamus*; *C. metzianus* and *C. rivalis* clustered together. Variability within populations is less than the variability between populations.	[8]
RAPD marker-aided diversity in *C. thwaitesii* in India and Sri Lanka	80 individuals of 5 populations of Western Ghats of India and 3 populations from Sri Lanka; 100 operon RAPD primers (OPA, OPB, OPE, OPAU, and OPAW)	10 RAPD primers (OPAU02, OPA02, OPA11, OPAW07, OPAW09, OPAW10, OPAW20, OPE01, OPE04, and OPB10) gave scorable bands. Highest genetic diversity measures were observed in Goa population. Variability within populations: variability is less than the variability between populations; Sri Lankan populations did not discriminate from Indian populations.	[9]
Locations type of conservation on rattan diversity using ISSR markers	25 ISSR primers from the UBC set (Sigma, USA) were screened	10 primers (UBC 890, UBC 841, UBC 835, UBC 834, UBC 868, UBC 880, UBC 855, UBC 848, ISSR 4, and ISSR 5) gave good results.	[11]

(*Continued*)

Table 5.1 (Continued)

Purpose	Materials and Methods	Conclusion	Reference
Species and genotype-specific RAPD bands in *Calamus Daemonorops* and *Plectocomia*	5 plants each in 15 different rattan genotypes, 13 belonging to the genus *Calamus* and 1 each of Daemonorops and Plectocomia, were collected from Assam and Arunachal Pradesh, India. 20 primers from OPD, OPE, and OPL operon series were tested; of these, 12 primers gave good results.	Identified genotype-specific markers suitable for discriminating 7 genotypes of *Calamus* and *D. jenkinsianus*.	[12]

analyzing the diversity of populations of *Calamus metzianus* and *C. rivalis* [8] and *C. thwaitesii* [9] in South India and Sri Lanka, North East Himalaya, a megadiversity region in India. DNA analysis indicated the populations of *C. thwaitesii* at Goa state in India as having high genetic diversity [9]. Molecular markers helped to identify the four diversity hot spot sites of the north-eastern Himalayas of India for conserving *C. flagellum* [10]. ISSR marker analysis was attempted on *C. thwaitesii* populations in three types of locations (protected areas, buffer zones, and peripheral areas) at three sites of Western ghats [11]. Protected areas conserve genetic resources of rattans better than the other two types of locations, indicating the vulnerability of rattans [11]. Species- and genotype-specific RAPD markers were discovered in *Calamus* [12] and need conversion as SCAR markers and further validation.

5.2.3 Marker–Sex Association in Rattans

A male-specific RAPD [13] marker S1443 (5′-TCTCGCCTAC-3′), with an amplification size of 500bp in *Calamus simplicifolius*, was converted to a SCAR marker and validated [14]. A female-specific ISSR marker (ISSR4—600 bp, (AAG)5CC– [15] was identified in *Calamus tenuis*.

5.3 Molecular Markers for Diversity and Sex Association in Palmyra

Amplification of DNA of palmyra palms was attempted with 16 operon OPZ series RAPD primers. Only seven primers (OPZ 01, OPZ 04, OPZ 05, OPZ 06, OPZ 07, OPZ 10, and OPZ12) gave scorable bands and were used to analyze a set of 20 individuals varying for sex, habit, and yield (four dwarf males, four dwarf females, four tall males, four tall females, two high neera yielding males, and two high neera yielding females). The dendrogram displayed groupings in four different clusters based on the stature, sex, and high neera yielding types [16]. Out of 180 RAPD primers screened, only three (OPBE-12, OPBA-13, and OPA-06) showed polymorphism between male and female sex pools of palmyra [17]. OPA-06 gave a consistent band of 600 bp [17] when the individuals were analyzed. A future priority in palmyra genomics would be the validation of these markers and development of SCAR markers to tag important traits.

References

1. Dransfield J, Uhl NW, Asmussen NCB, Baker WJ, Harley MM, Lewis CE. Genera palmarum: the evolution and classification of palms. Kew, UK: Royal Botanic Gardens; 2008.
2. Baldwin BG, Sanderson MJ, Porter JM, Wojciechowski MF, Campbell CS, Donoghue MJ. The ITS region of nuclear ribosomal DNA: a valuable source of evidence on angiosperm phylogeny. *Ann. Missouri Bot. Gard.* 1995;**82**:247–277.
3. Baker WJ, Hedderson TA, Dransfield J. Molecular phylogenetics of subfamily Calamoideae (Palmae) based on nrDNA ITS and cpDNA rps16 intron sequence data. *Mol. Phylogen. Evol.* 2000;**14**:195–217.
4. Barrow SC. Systematic studies in *Phoenix* L. (Palmae: Coyphoideae). *Mem. N. Y. Bot. Gard.* 1999;**83**:215–223.
5. Baker WJ, Hedderson TA, Dransfield J. Molecular phylogenetics of *Calamus* (palmae) and related rattan genera based on 5s nrDNA spacer sequence data. *Mol. Phylogenet. Evol.* 2000;**14**:218–231.
6. Teulat B, Aldam C, Trehin R, Lebrun P, Barker JH, Arnold GM, et al. An analysis of genetic diversity in coconut (*Cocos nucifera* L.) populations from across the geographic range using sequence-tagged microsatellites (SSRs) and AFLPs. *Theor. Appl. Genet.* 2000;**100**:764–771.
7. Rao NM, Ramesha BT, Ravikanth G, Ganeshaiah KN, Umashankar R. Cross-species amplification of coconut microsatellite markers in rattans. *Silvae Genet.* 2007:282–286.
8. Sreekumar VB, Renuka C, Suma TB, Balasundaram M. Taxonomic reconsideration of *Calamus rivalis* Thw. Ex Trim. and *C. metzianus* Schlecht (Arecaceae) through morphometric and molecular analyses. *Bot. Stud.* 2006;**47**:443–452.
9. Sreekumar VB, Renuka C. Assessment of genetic diversity in *Calamus thwaitesii* BECC (Arecaceae) using RAPD markers. *Biochem. Syst. Ecol.* 2006;**34**:397–405.
10. Lyngdoh N, Santhosh H, Ramesha BT, Nageshwara Rao M, Ravikanth G, Ganeshaiah KN, et al. Rattan species richness and population genetic structure of *Calamus flagellum* in North-Eastern Himalaya, India. *J. Bamboo Rattan.* 2005;**4**:293–307.
11. Ramesha BT, Ravikanth G, Nageshwara Rao M, Ganeshaiah KN, Uma Shaanker R. Genetic structure of rattan *Calamus thwaitesii* in core, buffer and peripheral regions of three protected areas at central Western Ghats, India: do protected areas serve as refugia for genetic resources of economically important plants? *J. Genet.* 2007;**86**:1–10.
12. Sarmah P, Barua PK, Sarma RN, Sen P, Deka PC. Genetic diversity among rattan genotypes from India based on RAPD-marker analysis. *Genet. Resour. Crop Evol.* 2007;**54**:593–600.

13. Yang H, Gan S-M, Yin G-T, Xu H-C. Identification of random amplified polymorphic DNA markers linked to sex determination in *Calamus simplicifolius*. *J. Integr. Plant Biol.* 2005;**47**:1249–1253.
14. Li M, Li F, Yang F, Yin G, Gan S. A male-specific SCAR marker in *Calamus simplicifolius*, a dioecious rattan species endemic to China. *Mol. Breed.* 2010;**25**:549–551.
15. Sarmah P, Sarma RN. Identification of a DNA marker linked to sex determination in *Calamus tenuis* Roxb., an economically important rattan species in northeast India. *Mol. Breed.* 2011;**27**:115–118.
16. Ponnuswami V, Jagadeesan R, Ramesh Kumar A, Prabhu M, Makesh S. Genetic relationship and diversity in palmyrah accessions based on RAPD markers. *Am. Eurasian J. Sustain. Agric.* 2008;**2**:165–171.
17. George J, Karun A, Manimekalai R, Rajesh MK, Remya P. Identification of RAPD markers linked to sex determination in palmyrah (*Borassus flabellifer* L.). *Curr. Sci.* 2007;**93**:1075–1076.

6 Arecanut

6.1 Diversity of *Areca*

There are very few reports available on molecular markers on arecanut. Moderate genetic diversity is observed in cultivated arecanut varieties, using 19 random amplified polymorphic DNA (RAPD) primers. Genetic distance [1] varied from 0.22 (between Sree Mangala and Maidan Local) to 0.47 (between Mohitnagar *inter se* and Mohitnagar). RAPD primers of OPM, OPAF, and OPC series were screened in arecanut [2] for testing clonal fidelity, where OPAF02 gave polymorphic bands when amplified at 42 °C for 1 min. Nine microsatellite loci (AC01, AC06, AC07, AC08, AC14, AC23, AC28, AC29, and AC30) were developed in areca [3]. All these loci differed in Hardy–Weinberg equilibrium except AC30. Linkage disequilibrium was seen only between AC06 and AC08 loci. Molecular markers linked to important traits of arecanut such as dwarf habit, high tannins, low arecoline, and arecaidine content are not yet available.

6.2 Molecular Phylogeny of *Areca*

Areca and seven other genera, *Gronophyllum*, *Gulubia*, *Hydriastele*, *Loxococcus*, *Nenga*, *Pinanga*, and *Siphokentia* [4], form the Arecinae subtribe (Figure 6.1). *Areca, Nenga*, and *Pinanga* form the areca clade, a major group in the subtribe (Figure 6.2). In this subtribe, intron sequences from phosphoribulokinase (PRK) and the second largest subunit of RNA polymerase II (RPB2) were employed to study the phylogenetic analysis. PRK intron showed an amplicon with a size of 700 base pair (bp) and a 50 bp variation among the members of the subtribe. *Areca* had a GC-rich homopolymer in PRK intron as a unique feature. RPB2 intron displayed an amplicon of 850 bp consistently across the Arecinae. These two DNA variations were useful in Arecinae to understand the relationships of the subtribe and its genera, and to examine the evolution of dichogamy, with different levels of protandry and protogyny [5]. Intron sizes of PRK phosphoribulokinase (PRK; a Calvin cycle enzyme) and malate synthase (MS; a glyoxylate cycle enzyme) showed amplicons of 900–1300 and 500 bp, respectively [6], in Areceae, the largest tribe of the Arecaceae family. DNA variations in these two low-copy nuclear loci were useful in constructing the phylogeny of the Areceae tribe with 17 subtribes and 54 species.

Genomics of Cultivated Palms. DOI: 10.1016/B978-0-12-387736-9.00006-6

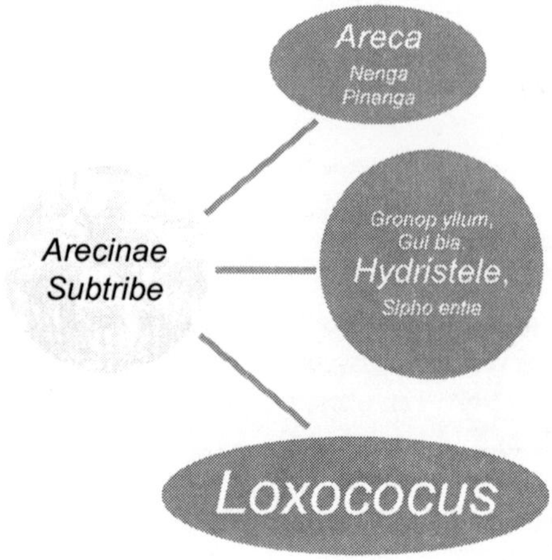

Figure 6.1 Phylogeny of Arecinae subtribe.

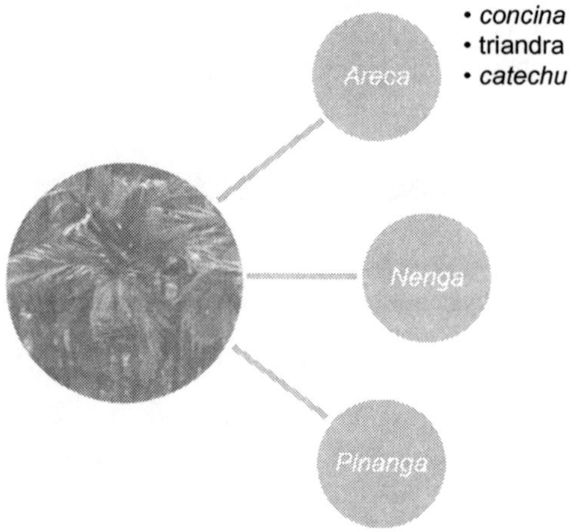

Figure 6.2 Phylogeny of areca clade of Arecinae subtribe.

6.3 Arecanut and Health Hazards

Arecanut, also called betel nut palm, is known as a mild stimulant and used by 200–400 million people [7] along with betel leaf (*Piper betle*), lime, and sometimes tobacco and spices. Arecanut is the fourth most (Figure 6.3) abused substance in the world [8], after nicotine (tobacco), alcohol, and caffeine, as well as a main cause of oral cancers. Arecanut-associated oral cancer or oral squamous cell carcinoma (OSCC) is the third most common malignancy in developing countries. Frequent consumption of areca causes a general fibrotic change in the cavity of oral submucosa and oral submucous fibrosis (OSF). It also leads to rigidity of oral and

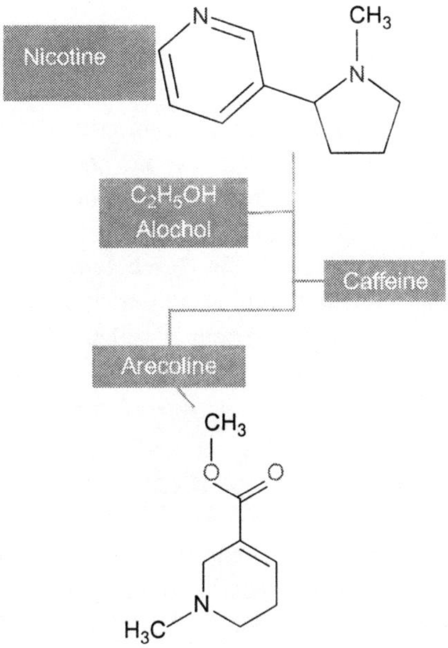

Figure 6.3 Most abused substances in the world.

lip tissues. The patient finds it difficult to open and close the mouth and move the tongue. The disease also makes the tongue sensitive to spicy food and causes rigidity of the lip [9]. Arecoline is the main alkaloid of arecanut, with a soluble feature in water and alcohol. During betel chewing, the concentration of arecoline [10] was reported to reach 140 μg/ml.

6.4 Biomarkers for Arecoline-Induced Damage

An understanding of the biochemical changes during areca consumption and the associated gene changes for oral cancer helps to identify the biomarkers associated with areca abuse. About 12 susceptible genes (five induced and seven repressed genes, including glutathione synthetase) can potentially serve as biomarkers of arecoline-induced damage in betel chewers. Expression changes of these genes were increased in a dose-dependent manner in a concentration range of arecoline at 50–150 μg/ml. To detoxify the arecanut and related ingredients, two monooxygenase systems (cytochrome P450 and flavin-containing monooxygenases) are considered important [11]. The free copper content was shown to high in arecanut ingredients [12], and increased soluble copper levels are found in saliva of chronic chewers. Copper acts as an initiating factor in oral carcinogenesis by stimulating fibrosis through up-regulation of lysyl oxidase enzyme activity [13].

The level of 8-hydroxydeoxyguanosine (8-OH-dG) in urine is an old known indicator of carcinogenesis; similarly, serum arecoline and arecaidine levels are proven to be good candidates for biomarkers for recent betel quid use [14]. Endogenous glutathione (GST) and the level of p52 protein are also proved as biomarkers for raw betel nut genotoxicity [15]. Arecoline-dependent avβ6 (alphavbeta6 integrin) up-regulation promoted keratinocyte migration and carcinogenesis. In more than 80% of oral cancers examined, there was moderate to high avβ6 expression [16]. Individuals having a long history of areca consumption sometimes do not develop oral cancers. Genetic polymorphism studies can be done to classify individuals according to their response to arecoline-induced carcinogenesis. Screening of patients for SNPs in the biomarker genes discussed above and correlating these data of carcinogenesis would help to understand which individuals or groups are likely to be not affected by areca consumption.

6.5 Xenobiosis, Metabolomics, and Pharmacogenomics

Pharmacogenomics is the study of genes and the gene products (proteins) essential for pharmacological or toxicological responses to pharmaceutical agents. Xenobiosis is a natural response to foreign compounds that elicits initiating signals for the many physiological events after pathogenesis. Metabolomics is a methodology for measuring small molecule metabolite levels and flux in biological matrices following genetic modifications or exogenous challenges. Metabolic maps are extensively investigated in other abused substances such as nicotine. Recently, metabolic maps of arecoline and arecaidine [17–19] were developed, which are helpful in understanding the clinical toxicology of areca alkaloids [17], especially 13 metabolites. These are mostly novel such as unchanged $\pm$ arecoline-1-oxide (50%), N-oxide metabolites (25%), and mercapturic acids. Arecoline on hydrolysis yields arecaidine that on nitro-oxidation forms mercaptauric acid [18]. Liquid chromatography–mass spectrometry (LC–MS), coupled with *in vivo* metabolism of several xenobiotics, is very efficient in identifying metabolites and elucidating *in vivo* metabolic pathways and overall metabolic maps [19]. Recently, the possible role of arecoline N-oxide in the induction of oral carcinogenesis by arecanut chewing was predicted due to its mutagenicity to the Salmonella tester strains [20]. A toxicogenomic [21] database for rANE (ripe arecanut extract) has been established using the down-regulation of KGF-1 expression data in oral fibroblast cell lines potentially impairing the proliferation of overlying keratinocytes. It partially explains the frequent epithelial atrophy observed in chronic areca chewers *in vivo*. Such database and resources offer scope in computational pharmacogenomic investigations using bioinformatics tools.

6.6 *Areca* Damage and Herbal Remedies

Arecanut has wide pharmacological properties as an antioxidant, anti-appetite, antibacterial, anti-venom, antihelmintic, and molluscicidal, hence must be appropriately used (review:23) at effective doses. *Areca* is a rich source of [22] tannins (8–18%)

and polyphenols. Arecanut leaf sheaths (43% cellulose) are used as paper boards [23], animal feed, and eco-friendly edible plates. Hence, banning of areca use can eliminate other beneficial uses of the plant. Chemopreventive interventions are being explored for individuals susceptible to areca-induced oral carcinogenesis. Retinoids and green tea are among the promising agents [24–26] that provide relief to patients. Although a large percentage of oral carcinogenic lesions did respond to such treatments, recurrence of the lesions after terminating the chemopreventive regime has been observed [27], perhaps due in part to continuation of the addictive habit. Computational metabolomics research is needed to identify the compounds that can provide effective long-term relief to areca addictive oral cancer patients.

6.7 Road Map for *Areca* Genomics

Areca is a unique palm with multiple medicinal uses but abused for arecoline only. Molecular markers (a limited number of RAPD and SSRs) are currently available. A dwarf mutant areca palm was seen in Hirehalli at South Canara District of Karnataka, India. It is a 40-year palm but attained only 4.57 m height and had compressed internodes with medium-size and slightly elongated nuts [28]. Dwarf palms have shorter stature (Figures 6.4 and 6.5) with compact internode than normal tall palms with long internodes (Figures 6.6 and 6.7). Central Plantation Crops Research Institute Regional Station (CPCRI RS) at Vittal South Canara District of Karnataka, India, maintains a large germplasm of the crop. The center also has developed many improved cultivars and engages in breeding dwarf, early bearing, and high-yielding hybrids. A large number of codominant markers have to be generated for characterizing diversity

Figure 6.4 Dwarf areca palm.

Figure 6.5 Internodes of dwarf areca.

Figure 6.6 Tall areca palm.

and linkage mapping, especially to tag dwarf trait (Figures 6.4 and 6.5), tannin content, arecoline content, and resistance to diseases such as bud rot and yellow leaf. Molecular phylogeny of areca and related taxa were worked out using MS, PRK, and RPB gene-intron sequence variations. WRKY, RGA (resistance gene analog), and protein kinase gene sequences must also be employed. Knowledge from metabolomics of areca alkaloids should complement the studies on chemopreventive substances. Research on biomarkers should be advanced for development of SNP chips and kits to locate humans susceptible to areca abuse. Alternate food additives have to be identified using this knowledge, to plan rehabilitation programs for areca abusers.

Figure 6.7 Internodes of tall areca palm.

References

1. Purushotham B, Narayanaswamy P, Simon L, Shyamalamma S, Hegde M, Jaypalgowdu B. Genetic relationship between cultivars of Arecanut (*Areca catechu* L.). *Asian Aust. J. Plant Sci. Biotech.* 2008;**2**:31–35.
2. Karun A, Radha E, Vijayan S, George J, Rajesh MK, Ananda KS. Assessment of genetic fidelity of arecanut plantlets derived by direct somatic embryogenesis by RAPD markers. *J. Plantation Crops.* 2008;**36**:228–231.
3. Hu CH, Huang CC, Hung KH, Hsu TW, Chiang TY. Isolation and characterization of polymorphic microsatellite loci from *Areca catechu* (Arecaceae) using PCR-based isolation of microsatellite arrays. *Mol. Ecol. Resour.* 2009;**9**:658–660.
4. Dransfield J, Uhl NW. An outline of a classification of palms. *Principes.* 1986;**30**:3–11.
5. Loo AHB, Dransfield J, Chase MW, Baker WJ. Low-copy nuclear DNA, phylogeny and the evolution of dichogamy in the betel nut palms and their relatives (Arecinae; Arecaceae). *Mol. Phylo. Evol.* 2006;**39**:598–618.
6. Lewis CE, Doyle JJ. A phylogenetic analysis of tribe Areceae (Arecaceae) using two low-copy nuclear genes. *Plant Syst. Evol.* 2002;**236**:1–17.
7. Gupta PC, Warnakulasuriya S. Global epidemiology of areca nut usage. *Addict. Biol.* 2002;**7**:77–83.
8. Norton SA. Betel: consumption and consequences. *J. Am. Acad. Dermatol.* 1998;**38**:81–88.
9. Tilakaratne WM, Klinikowski MF, Saku T, Peters TJ, Warnakulasuriya S. Oral submucous fibrosis: review on aetiology and pathogenesis. *Oral Oncol.* 2006;**42**:561–568.
10. Nair J, Ohshima H, Friesen M, Croisy A, Bhide SV, Bartsch H. Tobacco-specific and betel nut-specific N-nitroso compounds: occurrence in saliva and urine of betel quid chewers and formation *in vitro* by nitrosation of betel quid. *Carcinogenesis.* 1985;**6**:295–303.
11. Chiang SL, Jiang SS, Wang YJ, Chiang HC, Chen PH, Tu HP, et al. Characterization of arecoline-induced effects on cytotoxicity in normal human gingival fibroblasts by global gene expression profiling. *Toxicol. Sci.* 2007;**100**:66–74.
12. Raja KB, Hazarey VK, Peters TJ, Warnakulasuriya S. Effect of areca nut on salivary copper concentration in chronic chewers. *Biometals.* 2007;**20**:43–47.

13. Shieh TM, Tu HF, Ku TH, Chang SS, Chang KW, Liu CJ. Association between lysyl oxidase polymorphisms and oral submucous fibrosis in older male areca chewers. *J. Oral Pathol. Med.* 2009;**38**:109–113.
14. Wu IC, Chen PH, Wang CJ, Wu DC, Chyang D, Tsai SM, et al. Quantification of blood betel quid alkaloids and urinary 8-hydroxydeoxyguanosine in humans and their association with betel chewing habits. *J. Anal. Toxicol.* 2010;**34**:325–331.
15. Kumpawat K, Chatterjee A. The usefulness of cytogenetic parameters, level of p53 protein and endogenous glutathione as intermediate end-points in raw betel-nut genotoxicity. *Hum. Exp. Toxicol.* 2003;**22**:363–371.
16. Moutasim KA, Jenei V, Sapienza K, Marsh D, Weinreb PH, Violette SM, et al. Betel-derived alkaloid up-regulates keratinocyte alphavbeta6 integrin expression and promotes oral submucous fibrosis. *J. Pathol.* 2011;**223**:366–377.
17. Giri S, Krausz KW, Idle JR, Gonzalez FJ. The metabolomics of (+/–)-arecoline 1-oxide in the mouse and its formation by human flavin-containing monooxygenases. *Biochem. Pharmacol.* 2007;**73**:561–573.
18. Giri S, Idle JR, Chen C, Zabriskie TM, Krausz KW, Gonzalez FJ. A metabolomic approach to the metabolism of the areca nut alkaloids arecoline and arecaidine in the mouse. *Chem. Res. Toxicol.* 2006;**19**:818–827.
19. Chen C, Gonzalez FJ, Idle JR. LC-MS based metabolomics in drug metabolism. *Drug Metab. Rev.* 2007;**39**:581–597.
20. Lin KH, Lin CY, Liu CC, Chou MY, Lin JK. Arecoline N-oxide: its mutagenicity and possible role as ultimate carcinogen in areca oral carcinogenesis. *J. Agric. Food Chem.* 2011;**59**:3420–3428.
21. Ko SY, Lin SC, Chang KW, Liu CJ, Chang SS, Lu SY, et al. Modulation of KGF-1 gene expression in oral fibroblasts by ripe areca nut extract. *J. Oral Path. Med.* 2003;**32**:399–407.
22. Jaiswal P, Kumar P, Singh VK, Singh DK. *Areca catechu* L.: a valuable herbal medicine against different health problems. *Res. J. Med. Plant.* 2011;**5**:145–152.
23. Ragupathy R, Viswanathan R, Devadas CT. Quality of paper boards from arecanut leaf sheath. *Bioresour. Tech.* 2002;**82**:99–100.
24. Garewal H. Chemoprevention of oral cancer: beta-carotene and vitamin E in leukoplakia. *Eur. J. Cancer Prev.* 1994;**3**:101–107.
25. The Indian-US Head and Neck Cancer Cooperative Group (IUSHNCC) Green tea and leukoplakia. *Am. J. Surg.* 1997;**174**:552–555.
26. Papadimitrakopoulou VA, Hong WK. Retinoids in head and neck chemoprevention. *Proc. Soc. Exp. Biol. Med.* 1997;**216**:283–290.
27. Sankaranarayanan R, Mathew B, Varghese C. Chemoprevention of oral leukoplakia with vitamin A and beta carotene: an assessment. *Oral Oncol.* 1997;**33**:231–236.
28. Naidu GVB. Seen a dwarf areca palm? *Indian Farming.* 1963;**12**:16–17.

7 Peach Palm

7.1 Introduction

Peach palm, or Pejibaye or pupunhua (*Bactris gasipaes* Kunth), was the most important palm of pre-Columbian America. It was a main food crop of Amerindians and belongs to the subtribe Bactridinae with chromosome number $2n = 28$. Currently, the palm is commercially valuable as heart of palm and for its starchy, oily fruits. Peach palm has potential as an oil crop and is also rich in carotenes [1]. Information on the importance, uses, and botany of the palm is available at www.fao.org/docrep/t0646e/T0646E0l.htm. Molecular markers (allozymes, RAPD, AFLP, and SSRs) are used to analyze the diversity in wild and domesticated landraces of peach palm. Most of these reports found the lack of genetic differentiation of wild and cultivated landraces of peach palm. When the hearts (palmitos) of this palm are harvested, there is very little oxidation and they remain fresh. Polyphenol oxidase is the enzyme that readily causes oxidation of the tissues, which in turn causes browning. Peach palm does not express the gene for the polyphenol oxidase enzyme and hence offers quality [2]. Reviews [3–6] and dissertations [6,7] provide background information on the genetic diversity of peach palm. Molecular phylogeny studies on members of the Bactridinae subtribe, Bactris and its species, and marker studies on landraces, wild populations, and the report on polyphenol oxidase of peach palm are reviewed and discussed in this chapter.

7.2 Molecular Phylogeny of Bactridinae

The Bactridinae subtribe is part of the Cocoseae tribe of palms, consisting of five genera and 150 species of spiny Neotropical palms (Acrocomia, Astrocaryum, Bactris, Aiphanes, and Desmoncus). The genetic relationship of the 41 species of Bactridinae [8] was analyzed, using five plastid DNA regions (matK, trnQ-*rps*16, *rps*16 intron, trnD–trnT, trnL–trnF) along with three nuclear markers (PRK, RPB2, ITS). The study found *Desmoncus* and *Acrocomia* to be sister genera, and together seen as sister to the rest of the *Bactridinae* [8]. Two new single-copy nuclear genes, AGA1 and PHYB, were used for resolving the controversy of relationship of Bactridinae, especially *Astrocaryum* and *Hexopetion* [9]. The Hexopetion clade was found to be monophyletic in AGA1 and PHYB, but was not strongly supporting Astrocarym as a sister. Eiserhardt et al. [8] also found microsatellite repeat in these sequences, with potential as markers for taxonomy and population genetics [9]. Analysis of four chloroplast DNA regions, eight microsatellite loci on 38 individuals of 29 species of *Bactris* genus along with coconut, *Attlaeae crassipatha* [9], found

Genomics of Cultivated Palms. DOI: 10.1016/B978-0-12-387736-9.00007-8

the *Bactris* genus as monophyletic. Wild populations of *B. gasipaes* were found genetically similar to *B. riparia* from the study [10].

7.3 Biochemical and Molecular Markers in Peach Palm

Allozyme [11], RAPD (as cited in Refs. [3,12,13]), AFLP [14,15], and SSR [16–20] markers were used to understand the genetic diversity and relationship among landraces, germplasm accessions, and wild populations of peach palm. Table 7.1 lists the details of the results of the studies on biochemical and molecular marker-aided diversity in *Bactris*.

Bactris gasipaes is classified into two forms: *B. gasipaes var. gasipaes*, the domesticated type, and *B. gasipaes var. chichagui*, the ancestral wild type. Plants of wild and cultivated forms of *Bactris gasipaes* are considered as conspecific [21]. There are three types within the wild type, based on fruit size (21, as cited in Ref. [22]).

A support vector machine (SVM)-based algorithm and its software are capable of identifying the landraces of the peach palm, using data of 89 haplotypes of 12 RAPD primers [12]. Two AFLP primer combinations, EcoRI-AAC/MseI-CCC and EcoR1-ACA/MseI-CTG, gave more than 64 polymorphic bands each. Eighteen important microsatellite loci were developed at CIARD France, with cross-amplification ability in *Astrocarym* and *Elaeis* [19]. Another 18 SSR loci were developed [16] at CIAT, Columbia, for peach palm, some of which can also amplify other species of *Bactris*.

Microsatellite variation (Table 7.2) also aided in understanding genetic differentiation [15,17]; gene flow, origin, and migration routes [13]; genetic diversity and phylogenetic relationship among *Bactris* and its landraces; and wild types [17,18] and their hybrid intermediate forms [18]. High gene flow was seen among the landraces of Central America. But the movement of genes was low between the landraces of Amazon and Central America. Rodrigues, Filho, and Clement [13] also demonstrate the origin of southwestern Amazonian landraces that moved northeast either via a northeast direction (Para landrace) or northwest along the Andes through (Pampa Hermosa and Putumayo landraces) western Amazon up to Central America (Utilis landrace) [13]. South and north extremes of western Ecuador [17] are occupied by wild and cultivated Bactris forms, respectively [17]. Fruit production of cultivated plants is double that of wild plants; intermediate-sized fruits and mixed characteristics are noticed at sympatry of wild and cultivated in northwestern Ecuador [18]. SSR markers provided a method to interpret methods of sourcing seed, monitor changes in measures of genetic diversity in recipient populations over time, and scale the way the migration takes place [20]. Many wild relatives of several native South American crops are located in the Arc of Fire. Many peach palm populations suffer genetic erosion at the Arc of Fire by disturbances in the forests and could be saved by *in situ* conservation [22] strategies.

Table 7.1 Biochemical and Molecular Markers in Peach Palm

Purpose	Materials	Methods	Conclusion	Reference
Allozyme variation in spineless peach palm	10 open-pollinated progenies of 3 populations introduced to Hawaii	14 enzyme systems; 5 of 16 loci are monomorphic	Three populations are closely related; allozyme analysis is useful for fingerprinting.	[11]
Identification of landraces in peach palm	Data from RAPD experiments on 87 samples of landraces	SVM-based algorithm for 12 informative markers from 30 operon primers of OPA OPC series	All peach palm landraces could be identified by these RAPD profiles of 89 haplotypes and the algorithm.	[12]
RAPD-aided assessment of diversity and their phylogenetic relationships	220 individuals of 4 Amazonian and 3 Central American landraces	30 operon RAPD among primers—11 polymorphic; 7 (primers OPA-4, OPA-5, OPA-8, OPA-9, OPA-18, OPA-20, FC13) gave 113 markers, 97 of which were polymorphic	Pampa Hermosa landrace is distinct from the rest possibly due to selection against spines in the region.	[13]
AFLP-aided evaluation of relationships among putative landraces	99 plants representing 13 populations of 3 landraces along the Amazonas and Solimões rivers in Brazil	AFLP 6 primer combinations with 245 markers	Putumayo landrace extends along the Solimões River to Central Amazonia.	[14]
AFLP-aided genetic diversity of populations along the Paranapura and Cuiparillo Rivers, which are managed by indigenous and colonist farming communities, respectively.	402 palms—100 from nonindigenous, 302 from indigenous; 39 mother plants of 7 populations of Brazil; 1 population of Peru	80 EcoRI/MseI primer combinations; 2 good: EcoRI-AAC/MseI-CCC (78 polymorphic bands) and EcoR1-ACA/MseI-CTG (66 polymorphic bands)	Germplasm collecting strategies: minimum distance of 100 m between mother trees; sample from 30 to 50 trees per population.	[15]
Variation in varieties for nutritional and antioxidant properties	6 varities: Bolívia, Darién, Brazil, Costa Rica, Colombia, Guatuso	Carotenoid components and antioxidants	Colombia and Guatuso had a higher proportion of beta carotene; Guatuso also had high radical scavenging activity.	[23]

Table 7.2 Review of Microsatellite-Aided Diversity Assessment in Peach Palm

Purpose	Materials	Methods	Conclusion	Reference
Microsatellite loci for Cocoseae tribe	(GA)n repeat-enriched library	A (GA)n microsatellite-enriched library was constructed.	New set of 18 SSR loci found highly variable in Bactris and transferable to related species Astrocaryum, Elaeis genera	[16]
Genetic relationship of wild and cultivated populations in western Ecuador	83 wild and cultivated individuals (69—western Ecuador; 14—Amazonia and Central America)	Volumetric study of fruits; DNA data from 9 microsatellite loci (8 from the publication [16] mBgCIR010, [57, 58, 62, 71, 87, 94, 204] and one from wild type mBgCIR204)	Amazonian palms have 10 times higher fruit volume than wild ones; low genetic structure between the groups	[17]
Diversity of populations in wild and cultivated compartments, allopatry and sympatry conditions	83 wild palms of southwestern and northwestern Ecuador; 213 cultivated palms of Ecuador, French Guyana, and wild plants of eastern Andes	8 nuclear dinucleotidic microsatellite loci; 2 chloroplastic microsatellites	Low differentiation between wild and cultivated; extensive reciprocal introgression between wild and cultivated compartments in sympatry at northwestern Ecuador	[18]
Microsatellite markers; cross-amplification in 6 species of Bactris	14 individuals selected from a germplasm bank	18 polymorphic microsatellite markers found an average of 7 alleles per locus	Bg 1, 2, 6, 9, 10, 11, 14, 17, 20, 21, 24, 28, 44, 46, 51, 55, 56, 63 Bg 17 for all 6 species; Bg 1, 14, 21 no amplification in other species	[19]
Seed migration and swidden-fallow management in the genetic structure of the population in northeastern Peru	24 and 81 palms in 2 locations, 160 km apart—one occupied by indigenous Amerindians and the other by campesinos. 386 plant samples 37 samples of germplasm collection	28 known SSR microsatellite marker loci [19] and [16]; 3 best: Bg02-19, Bg02-4 Bg55	Maintenance of remnant individuals among the indigenous farmers and by sourcing seed outside the community balanced the inbreeding coefficient	[20]

7.4 Polyphenol Oxidase, Carotenoids, and Antioxidants in Bactris

Currently the palm is commercially valuable as heart of palm and for its starchy, oily fruits. Peach palm has potential as an oil crop and is also rich in carotenes [1]. It is an underutilized palm whose diversity is threatened. Polyphenol oxidase is the enzyme that readily causes oxidation of the tissues, which in turn causes browning. Two partial cDNA sequences were isolated [2] and sequenced from two species of Euterpe, Açaí and Juçara (EoPPO1 426bp, DQ123806). Semiquantitative RT-PCR expression assays showed that Açaí and Juçara PPOs were strongly expressed in palmitos and weakly expressed in leaves. Peach palm does not express the gene for the polyphenol oxidase enzyme (PPO; EC 1.14.18.1), and this is possibly why the palmetto/heart of palm shows no browning [2]. Variation in carotenoid type and composition and antioxidants among the six varieties [23] also could serve as a tool (Table 7.1) to identify cultivar and its nutritional importance.

Genomics and protemic applications have begun for this important Neotropical palm. In the near future, genomic tools will play an important role in conservation genetics, molecular breeding, QTL mapping, and product utilization and diversification of peach palm.

References

1. Arkcoll DB, Aguilar JPL. Peach palm (*Bactris gasipaes* H.B.K.), a new source of vegetable oil from the wet tropics. *J. Sci. Food Agric.* 1984;**35**:520–526.
2. Shimizu MM, Melo GA, Santos ABD, Bottcher A, Cesarino I, Araújo P, et al. (2011). Enzyme characterization, isolation and cDNA cloning of polyphenol oxidase in the hearts of palm of three commercially important species. *Plant Physiol. Biochem.* **49**:970–977.
3. Clement CR, Arajuo MDC, d'Eeckenbrugge GC, Pereira AA, Rodrigues DP. Origin and domestication of native Amazonian crops. *Diversity*. 2010;**2**:72–106.
4. Clement CR. Pejibaye (*Bactris gasipaes*). In: Smartt J, Simmonds NW, eds. *Evolution of crop plants* (2nd ed.). London: Longman; 1995, p. 383–388.
5. Mora-Urpí J, Weber JC, Clement CR. Peach palm. Bactris gasipaes Kunth Promoting the conservation and use of underutilized and neglected crops. 20. Rome: Institute of Plant Genetics and Crop Plant Research, Gatersleben and International Plant Genetic Resources Institute; 1997, p. 83.
6. Clement CR (1995). Growth and genetic analysis of pejibaye (*Bactris gasipaes* Kunth, Palmae) in Hawaii. Ph.D. Dissertation, University of Hawaii Manoa, Honolulu, 221 pp.
7. Cole DM (2004). Genetic diversity and population structure of peach palm (Bactris gasipaes Kunth) in agroforestry systems of the Peruvian Amazon. M.Sc. thesis, University of Florida, 67 pp. (http://etd.fcla.edu/UF/UFE0007363/cole_d.pdf).
8. Eiserhardt WL, Pintaud JC, Asmussen LC, Hahn WJ, Bernal R, Balslev H, et al. Phylogeny and divergence times of Bactridinae (Arecaceae, Palmae) based on plastid and nuclear DNA sequences. *Taxon*. 2011;**60**:485–498.
9. Ludeña B, Chabrillange N, Bertossi FA, Adam H, Tregear JW, and Pintaud JC (2011). Phylogenetic utility of the nuclear genes AGAMOUS 1 and PHYTOCHROME B in palms (Arecaceae): an example within Bactridinae. *Ann. Bot.* doi: 10.1093/aob/mcr191.

10. Couvreur TLP, Hahn WJ, de Granville J-J, Pham JL, Ludeña B, Pintaud J-C. Phylogenetic relationships of the cultivated palm *Bactris gasipaes* (Kunth) with its wild relatives inferred from non-coding chloroplastic sequences. *Syst. Bot.* 2007;**32**:519–530.
11. Clement CR, Mallikarjuna KA, Manshardt RM. Allozyme variation in spineless pejibaye (*Bactris gasipaes* Palmae). *Econ. Bot.* 1997;**51**:149–157.
12. Vasquez JL, Vasquez J, Briceno JC, Castillo E, Travieso CM. Feature selection of RAPD haplotypes for identifying peach palm (Bactris gasipaes) landraces using SVM. *WSEAS Trans. Computers*. 2010;**9**:205–214.
13. Rodrigues DP, Filho SA, Clement CR. Molecular marker mediated validation of morphologically defined landraces of pejibaye (*Bactris gasipaes* Kunth): and their phylogenetic relationships. *Genet. Resour. Crop Evol.* 2004;**81**:871–892.
14. Clement CR, Sousa NR, Rodrigues DP, Astolfi-Filho S, Moreno YN, Pascual VT, et al. Use of AFLPs to distinguish landraces of pejibaye (*Bactris gasipaes*) in Brazilian Amazonia. *Scientia Agricola*. 2002;**59**:749–753.
15. Adin A, Weber JC, Sotelo-Montes C, Vidaurre H, Vosman B, Smulders MJM. Genetic differentiation and trade among populations of peach palm (*Bactris gasipaes* Kunth) in the Peruvian Amazon—implications for genetic resource management. *Theor. Appl. Genet.* 2004;**108**:1564–1573.
16. Billotte N, Couvreur TLP, Marseillac N, Brottier P, Perthuis B, Vallejo M, et al. A new set of microsatellite markers for the peach palm (*Bactris gasipaes* Kunth): characterization and across-taxa utility within the tribe Cocoeae. *Mol. Ecol. Notes*. 2004;**4**:580–582.
17. Couvreur TLP, Billotte N, Risterucci AM, Lara C, Vigouroux Y, Luden B, et al. Close genetic proximity between cultivated and wild *Bactris gasipaes* Kunth revealed by microsatellite markers in Western Ecuador. *Genet. Resour. Crop Evol.* 2006;**53**:1361–1373.
18. Pintaud JC, Couvreur TLP, Billotte N, Ludena B, Lara C, Pham JL. Structure and dynamics of genetic diversity in a wild and cultivated species complex: the case of the South-American Palm (*Bactris gasipaes*). *Les Actes du BRG*. 2006;**6**:355–370.
19. Martınez AK, Gaitan-Solis E, Duque MC, Bernal R, Tohme J. Microsatellite loci in Bactris gasipaes (Arecaceae): their isolation and characterization. *Mol. Ecol. Notes*. 2002;**2**:408–410.
20. Cole DM, White TL, Nair PKR. Maintaining genetic resources of peach palm (Bactris gasipaes Kunth): the role of seed migration and swidden-fallow management in northeastern Peru. *Genet. Resour. Crop Evol.* 2007;**54**:189–204, 10.1007/s10722-005-3134-3.
21. Henderson, A. (2000). Bactris (Palmae). New York: Flora Neotropica Monograph 79. New York: New York Botanical Garden.
22. Clement CR, Santos RP, Desmouliere SJM, Ferreira EJL, Neto JTF. Ecological adaptation of wild peach palm, its in situ conservation and deforestation-mediated extinction in Southern Brazilian Amazonia. PLoS ONE. 2009;**4**:e4564, 10.1371/journal.pone.0004564.
23. Jatunov S, Queseda S, Diaz. C, Murillo E. Carotenoid composition and antioxidant activity of the raw and boiled fruit mesocarp of six varieties of Bactris gasipaes. *Archivos latinoamericanos de nutricion Organo Oficial de la Sociedad Latinoamericana de Nutrición*. 2010;**60**:99–104.

8 Ornamental, Underutilized, and Vulnerable Palms

8.1 Attalea, Butia, Carpentaria and Ceroxylon

Attalea is an important palm genus of Brazil, Peru, and Paraguay. Piassava fibers, extracted from the leaf bases of *A. funifera*, provide a good income to Brazilian farmers. *Attalea phalerata*, commonly known as Scheelea palm or Shapaja, is a short palm, reaching less than 4 m in height, whose leaves are used as thatch and whose fibrous seed coating is eaten when ripe. Microsatellite markers [1] are useful in the study of the dispersal and spatial genetic structure of the *A. phalerata* palm populations. The markers also serve in analyzing population genetics and parentage. *Butia*, commonly known as Pindo palm, is native to South America in Brazil, Paraguay, Uruguay, and Argentina. Most species produce edible fruits that are sometimes used to make alcoholic drinks. ISSR markers [2] reveal the interspecific hybridization and gene flow among species. Cross-transferrable SSR markers [3] have been demonstrated by Nazareno, Zucchi, and Dos Reis in assessing vulnerable species of *Butia*. Molecular markers reveal the gene flow, as reported by researchers, among palm populations. Gene flow was low in *Carpentaria* [4] and *Ceroxylon* [5]. Details of the above discussions are given in Table 8.1.

8.2 Chamaedorea and Euterpe

Chamaedorea ernesti-augusti, also called Fishtail palm, is a slow-growing small understory palm. The leaves of the Fishtail palm are long and attractive and hence best suited for use as ornamental foliage. They are collected from forests and used in the cut flower industry. AFLP [6]; PRK and RPB2 [7]; plastid regions, PRK, and RPB2 [8]; SSR loci [9]; and allozymes [10] were used to understand the phylogeny in the Chamaedoreae tribe. SSR loci revealed the inbred nature of the species and identified the spatial genetic structure of Fishtail palm. Cultivar Tamash has high diversity and Columbia has limited allelic diversity. Genetic patterns reflect two major landscape events during the early Holocene (10,000 B.P. (Before Present)) and aridification during (1000) B.P. [9] in the species *Chamaedorea ernesti-augusti*. Heart of palm, also known as "palmito" in Brazil, is composed of the apical meristem of the palm plus part of the immature leaves of the meristem. It is used in salads and soups and hence exported. Two species of *Euterpe E. olearcea* (Açaí) and *E. edulis* (Juçara) are cultivated for heart of palms. Allozymes [11], SSR [12], and AFLP [13] markers are used to analyze the genetic diversity of Euterpe. Shimizu et al. [14] isolated,

Genomics of Cultivated Palms. DOI: 10.1016/B978-0-12-387736-9.00008-X

Table 8.1 Genomics of Attalea, Butia, Carpentaria, and Ceroxylon

Genus and System	Methods and Conclusion	Reference
Attalea—SSR	*A. phalerata* 14 microsatellite loci; 10 showed no deviations from Hardy-Weinberg equilibrium	[1]
Butia, Syagrus—ISSR	8 populations of 4 species of Butia-5 anchor ISSR primers $(CTC)_4RT$, $(AG)_8YT$, $(AC)_8T$, $(CA)_8G$, and $(GA)_8T$; low variability—evidence of hybridization, shared ancestory, and gene flow between species	[2]
Butia—SSR	50 samples of 2 populations of *B. eriospatha* (Mart. ex Drude) Becc analyzed using 14 microsatellite loci of GA/CA repeats; most of these loci are cross-transferrable to another vulnerable species, *B. catarinensis.* Noblick & Lorenzi	[3]
Carpentaria—Isozyme	High degree of genetic heterogeneity among *C. acuminata* populations; gene flow among populations was low; rainforest contraction during pleiocene	[4]
Ceroxylon—landscape genetics	*Ceroxylon echinulatum* geographical cline species status of Andes region confirmed by fruit morphology; gene flow limited by distance and by climatic friction; historical isolation in southwestern population	[5]

cloned, and sequenced two partial cDNA sequences of polyphenol oxidase gene from Açaí (EoPPO1) and Juçara (EePPO1). Gene expression assays indicated these to be strongly expressed in palmitos but weakly expressed in leaves [14]. In analyzing *Euterpe*, Panza et al. [15] found a dehydrin-like protein in seeds [15]. Storage proteins similar to those in *Phoenix*, useful as protein markers in palms, have also been reported [16]. Details of the above discussions are given in Table 8.2.

8.3 Geonoma, Howea, Johannesteijsma, Livistona, Licuala, Lodoicea and Metroxylon

The *Geonoma* genus has 68 species and is found in the tropical Americas. Its diversification occurred during the Miocene, 11.9–19.5 million years ago (mya) [17]. The degree of geographic structure in the *Geonoma* phylogeny has been investigated [17], and a significant geographic phylogenetic structure was found. Current patterns of geographic diversity in *Geonoma* reflect the present-day climate but to a larger extent are related to nonenvironmental spatial constraints linked to colonization time and dispersal limitation [17]. Sympatric varieties were found only in the northern region. There was no correlation between geographical distance and genetic proximity of individuals [18]. ISSR markers [18] and PRK and RPB2 sequence variations [19] are also used in *Geonoma*. Details of the above discussions are listed in Table 8.3.

Table 8.2 Genomics of Chamaedorea and Euterpe

Genus and System	Methods and Conclusion	Reference
Chamaedorea—AFLP	249 samples (127—*C. tepejilote* Lieb.; 122—*C. alternans* Wendl.) of 44 accessions in each using primer combinations EcoRI -GAG TGC GTA CCA ATT C- & MseI; GAT GAG TCC TGA GTA A; 2 species are genetically different; 5 of 45 loci display species specific fragments	[6]
Chamaedorea—PRK, RPB2	Sequence data of PRK (700 bp) and RPB2 (800 bp) from 63 species from 8 subgenera, *Chamaedorea* (57 species) confirmed as monophyletic genus; a 6 bp C-rich repeat, poly-T (22 bp) and poly-C (62 bp) regions of RPB2 interfered PCR amplification in 7 species; -A homopolymer (31 bp) in PRK prevented alignment across species.	[7]
Chamaedorea tribe-PRK RPB2, trnL–trnF, *rps*16, ndhF, matK, and trnD–trnT	40 species of tribe Chamaedoreeae—Hyophorbe, Gaussia, Synechanthus, and Wendlandiella-all species; Chamaedorea—27 species; PRK (531–638 bp), RPB2 (666–828 bp), and 5 plastid regions (trnL–trnF, *rps*16, ndhF, matK, and trnD–trnT); biography—northern distribution; 5 loci (Plastid regions—127 informative sites; RPB2-141 informative sites) good for phylogenetic classification	[8]
Chamaedorea—SSR	138 individuals of *C. ernesti-augusti* in 8 locations of Belize, Central America by 9 SSR loci; high degree of inbreeding, linkage disequilibrium at few locations	[9]
Chamaedorea—Allozyme	126 individuals of *C. tepejilote* (59 alleles) and 183 of *C. elatior* (53 alleles) palms; higher spatial genetic structure in *C. tepejilote* than *C. elatior*; because of diverse mix of family patches with many individuals; Hence (high degree of within population) genetic diversity	[10]
Euterpe—Allozyme	70–100 individuals of *E. edulis* in 6 plots at Santa Catarina (Brazil); 10 enzyme systems gave 11 ploymorphic loci-peroxidase, 6-phosphogluconate dehydrogenase and glucose-6-phosphate dehydrogenase gave specific alleles linked to recruitment process	[11]
Euterpe—SSR	66 individuals in natural population at Cerrado gallery of Atlantic using 18 polymorphic SSRs helpful in gene flow analysis & paternity testing	[12]
Euterpe—AFLP	150 plants of 11 populations analyzed by 429 AFLP markers reveal moderate within population variation (57%); geographical distance and genetic distance are positively correlated in these populations	[13]

Table 8.3 Genomics of Geonoma, Howea, Johannesteijsmannia, Livistona, Licuala, Lodoicea and Metroxylon

Genus and System	Methods and Conclusion	Reference
Geonoma genus PRK RPB2	11 taxa of Central America, 4 from French Guiana, 1 from Bolívia, 1 of Peruvian Andes analyzed sequence variations of intron 4 of PRK and intron 23 of RPB2 and range map data and 12 environmental factors for all Geonoma species	[17]
Geonoma—ISSR	31 individuals of 2 varieties of *G. macrostachys* at 4 locations of western Amazon (acaulis, macrostachys); 27 ISSR primers—99 bands-52 polymorphic; 2 main groups seen by geographical provenance	[18]
Geonomae tribe—PRK & RPB2	46 species of 6 genera of geonomae, 14 of Arecoideae, 2 of Ceroxyloideae, 1 of Phytelephantoideae analysed by PRK, RPB2 sequence variations; Geonoma is monophyletic with 2 main clades	[19]
Howea—AFLP	48 primer pairs initially and pairs B13, G7, and Y1 gave 274 AFLP bands on 2 species; high degree of homogeneity in 2 species	[20]
Johannesteijsmannia—SSR	24 samples *of J. lanceolata* using 31 polymorphic microsatellite markers helped understand mating system, heterozygosity. These markers are also cross transferrable to *J. altifrons, J. magnifica,* and *J. perakensis*	[21]
Livistona	*L. carinensis* is a vulnerable species in Africa limited to yemen, Djibouri, and Somalia. Nine SSR loci screened in 3 populations of Djibouri, Yemen, and Somalia. Djibouri monomorphic across all loci; other populations in botanic gardens as few individuals display heterozygosity and diverse than Djibouri	[22]
Livistona & Licuala-six-locus cpDNA data, MS intron	18 Australian Livistona species & other taxa across tribe Trachycarpeae analyzed by 1 nDNA (malate synthase, MS) and 6 cpDNA loci (total of 4000kb); Livistona is a monophyletic genus and sister to Licuala grandis based on 6 cp loci psbB–psbF, rpl20–*rps*12, trnT–trnD, trnS-P–trnfM-M (psbZ), trnL intron, and trnL–trnF exons and intergenic spacer, and MS; MS proved as a fast evolving locus in Livistona	[23]
Lodoicea—AFLP	180 trees of *L. maldivica* at 1 population each of 2 islands (Praslin and Curieuse) of Seychelles Indian Ocean; weak genetic differentiation between 2 populations	[24]
Metroxylon—AFLP	76 accessions of *M. sagu*, collected in 7 wild and semi-wild stands in Papau New Guinea; showed a significant correlation between genetic and geographical distances	[25]

Lord Howe is an oceanic island in the Pacific region between Australia and New Zealand. The ancestral form of *Howea* has existed for 4.5–5.5 myr (million years) at Lord Howe Island; but *H. forsteriana* recently diverged from the species *H. belmoreana* by colonizing in the calcarenite soil [20]. Both *Howea* species are highly homogenous and display very low genetic diversity. With the help of AFLP markers, Savolainen et al. explain the sympatric speciation in *Howea* [20]. *Johannesteijsmannia lanceolata* and a few other species of the genus are endangered palms. Cross-transferrable SSR markers developed in the genus are helpful in assessing heterozygosity and population genetics of other species of the genus [21]. *Livistona carinensis* is a vulnerable species limited to three locations in Africa. Nine SSR markers screened in the species gave the diversity status of the populations and their conservation priorities [22]. The geographical distribution of *Livistona* genus on both sides of the Wallace Line near Australia provides interesting material for molecular phylogeny [23]. Based on six chloroplast DNA regions and one nuclear DNA region, *Livistona mariae* and *L. rigida* were found to be identical across all loci sequenced (Table 8.3).

Double coconut, also called coco de mer (*Lodoicea maldivica*), is the largest-bearing dioecious palm in the entire plant kingdom. A very high (99%) degree of within-population AFLP diversity is seen in two island populations of the species, and little genetic differentiation between populations [24]. Coco de mer also displays high phenotypic plasticity and hence Dogley et al. suggest avoiding transferring seeds between islands to foster diversity [24]. AFLP markers indicate the correlation between genetic and geographical distance in sago palm [25] populations of Papua New Guinea (Table 8.3).

8.4 Nypa, Oenocarpus, Pinanga, and Pseudophoenix

Nypa fruitcans is an interesting palm belonging to monophyletic Nypoideae subfamily and found in mangrove ecosystems. SSR and ISSR marker analyses indicated the low genetic diversity in the species [26]. RAPD, ISSR markers [27], and PCR-RFLP of plastid DNA regions [28] were employed by Parani et al. to compare the *Nypa* with other plant species in mangrove ecosystem (Table 8.4). *Oenocarpus bacaba* is subcanopy arboreal palm distributed in the Amazon region that produces edible fruits liked by birds and animals. The species also provides nutritional fruit juice and oil [29]. SSR markers developed for the crop help in understanding the mating system and paternity analysis [29]. *Pinanga* is a large genus and displays variation in size and form [30]. Molecular phylogeny using the ITS region [30] of the ribosomal gene indicates the monophyletic nature of the *Pinanga* genus. RAPD markers in Indonesian populations of *P. javanica* indicate the genetic differentiation due to hybridization among the six sympatric populations. Hence Witona, Masuda, and Kondo [31] suggested both *in situ* and *ex situ* conservation of *Pinanga*. Genetic diversity of *Pinanga* is correlated with population density (Table 8.4). Gene flow among populations was high among the common species and low among the rare ones [32]. Interestingly, Shapcott found all the species of *Pinanga* to be inbred but genetically diverse [32] (Table 8.4).

Table 8.4 Genomics of *Nypa*, *Oenocarpus*, *Pinanga*, and *Pseudophoenix*

Genus and System	Methods and Conclusion	Reference
Nypa—RFLP, RAPD, ISSR	E*co*RI, *Hin*dII based RFLPs; 20 RAPDs from OPA, D, AR AF series, 3 microsatellites as primer (GATA)4, (GACA)4 & (GTG)6; of 23 primers only 9 RAPDs and (GATA)4 gave distinct bands in *Nypa* and other 21 mangrove species	[27]
Nypa—*trn*S-*psb*C, *rbc*L + *Hae* III	Primers for *trn*S-*psb*C (1600 bp) and *rbc*L gene (1400 bp) regions followed by digestion by *Hae*III gave unique banding pattern in *Nypa* compared to other dicot mangrove species	[28]
Nypa—SSR, ISSR	183 individuals of 6 populations of *N. fruticans* separated by 10–1200 km at Southeast Asia; 17 SSRs—2 polymorphic; 40 UBC ISSR primers gave 104 bands—3 polymorphic between populations	[26]
Oenocarpus—SSR	Eight SSR loci (AY 262720–AY 262728) of Oencocarpus bacaba and 2 of Euterpe; 5 did not fit Hardy-Weinberg equilibrium; markers useful to estimate mating system and analyze paternity	[29]
Pinanga—ITS rDNA	52 species amplified and sequenced (AB 271305–AB 271428) 3 regions (ITS1, ITS2, 5.8S), genus *Pinanga* is monphyletic	[30]
Pinanga—RAPD	88 accessions of *P. javanica* in 6 populations of java, 40 operon primers OPA and OPB series 10 (OPA—7,10,13,18, 20; OPB—2,7,8,11,18) found good	[31]
Pritchardia—SSR	6 new SSR loci and 3 known from Phoenix tested on Hawaiian species; 5 loci were polymorphic, with 72 alleles in Pritchardia martii	[33]
Pseudophoenix—SSR	104 individuals of 4 populations (3 mainland, 1 island) of *P. ekmanii* of Dominican Republic using 7 SSR loci revealed genetic drift, inbreeding, and moderate gene flow	[34]

Pritchardia (Loulu palm) is an important palm genus of Hawaii that has many vulnerable species. Five SSR loci developed in *P. marti* are cross-transferrable to 27 other species of the genus. These markers are useful in estimating species delimitation and intraspecific hybridization and planning conservation of *Pritchardia* [33]. *Pseudophoenix ekmanii* is another fragile and threatened palm species endemic to the Dominican Republic. Sap is extracted from the trees to make a local drink, and the tree dies after tapping. Using 7 SSR loci, Namoff et al. [34] found inbreeding and moderate gene flow, and also data that confirm an isolation-by-distance model. They found two groups, one from the north and another from the south. Lowest genetic diversity was observed in the island population [34] of *Pseudophoenix* (Table 8.4).

8.5 Conclusions

Molecular markers reports available for ornamental, underutilized, and vulnerable palms give interesting conclusions. Scientists have used a range of molecular markers, microsatellite loci, ISSR, AFLP, ITS, and RAPD in analyzing isolation by distance, species limits, gene flow, parentage, population genetics, and spatial genetic structure of these species. Genomics confirmed monophyly of three genera: *Chamaedorea*, *Geonoma*, and *Livistona*. We reviewed and discussed the cross-transferrable markers across species in these genera: AFLP in *Chamaedorea* [6] and SSR in *Butia* [3], *Pritchardia* [33], and *Johannesteijsmannia* [21]. Gene flow between populations is high in *Butia* [2] but moderate in *Pseudophoenix* [34] and low in *Carpentaria* [4] and *Ceroxylon* [5]. Positive correlation between geographical distance and genetic distance was found in *Euterpe* (AFLP [13]) and Sago palm (AFLP [25]), but no such correlation was seen in *Geonoma* (ISSR [18]). Genetic diversity was high in *Pinanga* [31] but low in *Nypa* [26]. A high degree of inbreeding is reported in Fishtail palm [9], Loulu palm [34], and Indonesian species of *Pinanga* [32]. Genomics is useful in estimating historical and climatic events in species evolution and in assessing diversity and interbreeding with other species, and hence is a great tool for conservation genetics.

References

1. Choo J, Ishak H, Simpson B, Mueller U, Juenger T. Characterization of 14 microsatellite loci in a tropical palm, Attalea phalerata (Arecaceae). *Am. J. Bot.* 2010;**97**:e105–e106.
2. Gaiero P, Mazzella C, Agostini G, Bertolazzi S, Rossato M. Genetic diversity among endangered Uruguayan populations of Butia Becc. species based on ISSR. *Plant Syst. Evol.* 2011;**292**:105–116.
3. Nazareno AG, Zucchi MI, and Dos Reis MS (2011). Microsatellite markers for Butia eriospatha (Arecaceae), a vulnerable palm species from the Atlantic Rainforest of Brazil. *Am. J. Bot.* **98**:e198–e200.
4. Shapcott A. The patterns of genetic diversity in Carpentaria acuminata (Arecaceae), and rainforest history in northern Australia. *Mol. Ecol.* 1998;**7**:833–847.
5. Trenel P, Gustafsson MHG, Baker WJ, Asmussen-Lange CB, Dransfield J, Borchsenius F. Mid-tertiary dispersal, not Gondwanan vicariance explains distribution patterns in the wax palm subfamily (Ceroxyloideae: Arecaceae). *Mol. Phylogenet. Evol.* 2007;**45**:272–288.
6. Bacon CD, Bailey CD. Taxonomy and conservation: a case study from chamaedorea alternans. *Ann. Bot.* 2006;**98**:755–763, 10.1093/aob/mcl158.
7. Thomas MM, Garwood NC, Baker WJ, Henderson SA, Russell SJ, Hodel DR, et al. Molecular phylogeny of the palm genus Chamaedorea, based on the low-copy nuclear genes PRK and RPB2. *Mol. Phylogenet. Evol.* 2006;**38**:398–415.
8. Cuenca A, Asmussen-Lange CB, Borchsenius F. A dated phylogeny of the palm tribe Chamaedoreeae supports Eocene dispersal between Africa, North and South America. *Mol. Phylogenet. Evol.* 2008;**46**:760–775.
9. Jaramillo AC, Bacon CD, Garwood NC, Bateman RM, Thomas MM, Russell S, et al. Population genetics of the understory fishtail palm Chamaedorea ernesti-augusti in Belize: high genetic connectivity with local differentiation. *BMC Genet.* 2009;**10**:65.

10. Luna R, Epperson BK, Oyama K. Spatial genetic structure of two sympatric neotropical palms with contrasting life histories. *Heredity*. 2005;**95**:298–305.
11. Conte R, Nodari RO, Vencovsky R, Reis MS. Genetic diversity and recruitment of the tropical palm, *Euterpe edulis* Mart., in a natural population from the Brazilian Atlantic Forest. *Heredity*. 2003;**91**:401–406.
12. Gaiotto FA, Grattapaglia D, Vencovsky R. Genetic structure, mating system, and long-distance gene flow in heart of palm (*Euterpe edulis* Mart.). *J. Hered.* 2003;**94**:399–406.
13. Cardoso SRS, Eloy NB, Provan J, Cardoso MA, Ferreira PCG. Genetic differentiation of *Euterpe edulis* Mart. populations estimated by AFLP analysis. *Mol. Ecol.* 2000;**9**:1753–1760.
14. Shimizu MM, Melo GA, Santos ABD, Bottcher A, Cesarino I, Araújo P, et al. (2011). Enzyme characterization, isolation and cDNA cloning of polyphenol oxidase in the hearts of palm of three commercially important species. *Plant Physiol. Biochem.* **49**:970–977.
15. Panza V, Distéfano AJ, Carjuzaa P, Láinez V, Vas MD, Maldonado S. Detection of dehydrin-like proteins in embryos and endosperm of mature Euterpe edulis seeds. Protoplasma. 2007;**231**:1–5, 10.1007/s00709-007-0248-9.
16. Panza V, Pighin D, Láinez V, Pollero RJ, Maldonado S. Storage lipids and proteins of *Euterpe edulis* seeds. *Biocell.* 2009;**33**:99–106.
17. Roncal J, Overgaard AB, Borchsenius F, Balslev H, Svenning JC. A dated phylogeny complements macroecological analysis to explain the diversity patterns in Geonoma (Arecaceae). *Biotropica.* 2011;**43**:324–334.
18. Roncal J, Ortega JF, Lewis CE. An evaluation of the taxonomic distinctness of two *Geonoma macrostachys* (Arecaceae) varieties based on intersimple sequence repeat (ISSR) variation. *Bot. J. Linn. Soc.* 2007;**153**:381–392.
19. Roncal J, Ortega JF, Asmussen CB, Lewis CE. Molecular phylogenetics of tribe Geonomeae (Arecaceae) using nuclear DNA sequences of phosphoribulokinase and RNA polymerase II. *Syst. Bot.* 2005;**30**:275–283.
20. Savolainen V, Anstett M, Lexer C, Hutton I, Clarkson JJ, Norup MV, et al. Sympatric speciation in palms on an oceanic island. *Nature*. 2006;**441**:210–213.
21. Ang CC, Lee SL, Lee CT, Tnah LH, Zakaria RM, Ng CC. Isolation and characterization of microsatellite loci in an endangered palm, Johannesteijsmannia lanceolata (Arecaceae). *Am. J. Bot.* 2011;**98**:e117–e119.
22. Shapcott A, Dowe JL, Ford DH. Low genetic diversity and recovery implications of the vulnerable Bankouale palm Livistona cariensis from north east Africa and the southern Arabian peninsula. *Conserv. Genet.* 2009;**10**:317–327.
23. Crisp MD, Isagi Y, Kato Y, Cook LG, Bowman DMJS. Livistona palms in Australia: ancient relics or opportunistic immigrants? *Mol. Phylogenet. Evol.* 2010;**54**:512–523.
24. Dogley FF, Kettle CJ, Edwards PJ, Ghazoul J, Maattanen K, Kaiser-Buunbury CN. Morphological and genetic differentiation in populations of the dispersal-limited coco de mer (*Lodoicea maldivica*): implications for management and conservation. *Divers. Distribut.* 2010;**17**:235–245.
25. Kjaer A, Barfod AS, Asmussen CB, Seberg O. Investigation of genetic and morphological variation in the sago palm (*Metroxylon sagu*; Arecaceae) in Papua New Guinea. *Ann. Bot.* 2004;**94**:109–117.
26. Jian S, Ban J, Ren H, Yan H. Low genetic variation detected within the widespread mangrove species Nypa fruticans (Palmae) from Southeast Asia. *Aquat. Bot.* 2010;**92**:23–27, 10.1016/j.aquabot.2009.09.003.

27. Parani M, Lakshmi M, Senthilkumar P, Ram N, Parida A. Molecular phylogeny of mangroves V. Analysis of genome relationships in mangrove species using RAPD and RFLP markers. *Theor. Appl. Genet.* 1998;**97**:617–625.
28. Parani M, Lakshmi M, Ziegenhagen B, Fladung M, Senthilkumar P, Parida A. Molecular phylogeny of mangroves VII. PCR-RFLP of trnS-psbC and rbcL gene regions in 24 mangrove and mangrove-associate species. *Theor. Appl. Genet.* 2000;**100**:454–460.
29. Cunha NL, Lund CA, Hamilton MB. Isolation and characterisation of nuclear microsatellite loci in the tropical arboreal palm *Oenocarpus bacaba* (Arecaceae). *Mol. Ecol. Notes.* 2003;**3**:435–437.
30. Witono JR, Kondo K. Molecular phylogeny of Pinanga based on internally transcribed spacer (ITS) sequence data. *Chromos. Bot.* 2007;**2**:25–37.
31. Witono JR, Masuda Y, Kondo K. Genetic diversity of Pinanga javana Blume in six natural populations in Java, Indonesia. *Chromos. Bot.* 2006;**1**:33–39.
32. Shapcott A. Comparison of the population genetics and densities of five Pinanga palm species at Kuala Belalong, Brunei. *Mol. Ecol.* 1999;**8**:1641–1654.
33. Bacon CD, Johnson GP, Meimberg H, Puppo P, Simmons MP, Wagner WL. Development of microsatellites in the Hawaiian endemic palm *Pritchardia martii* (Arecaceae) and their utility in congeners. *Am. J. Bot.* 2011;**98**:139–140.
34. Namoff S, Veloz A, Jiménez F, Rodríguez-Peña RA, Peguero B, Lewis C, et al. Sweet drinks are made of this: conservation genetics of an endemic palm species from the Dominican Republic. *J. Hered.* 2010;**102**:1–10.

9 Molecular Phylogeny of Palms

9.1 Introduction

We have thus far covered the use of genomics in individual palm species or genus. We devote this chapter exclusively to the use of DNA sequence variations in rearranging to clarify the taxonomic relationship of palms. A taxon (plural—taxa) is any group of organisms that is given a formal taxonomic name. A taxon is called monophyletic if it includes a group of organisms descending from a single ancestor or a common ancestor. If some of the members of the taxa do not share a common ancestor, then it is known as paraphyletic. A polyphyletic taxon is one whose members do not all have a common ancestor. Arecaceae, the palm family, is recognized as a monophyletic group among plants. The Arecaceae family is located within the commelinid clade of the monocotyledons group of plants. Arecoideae, Calamoideae, Ceroxyloideae, Coryphoideae, and Nypoideae are the five subfamilies of the palm family. Presently, the relationship among the 183 genera and 2364 species [1,2] of these five subfamilies is not completely resolved. A growing knowledge on the subject has made inroads toward achieving the objective. In estimating the phylogenetic structure of a taxon, the strength of DNA markers is measured by the number of parsimony informative (PI) sites or characters.

9.2 Random Primers and DNA Barcodes [2–5]

RAPD and ISTR genetic markers [3] identified and differentiated 66 genotypes in 50 palm genera. However, these markers are not suitable for genetic relationship analysis above the genus level [3]. A three-locus DNA barcode was used to identify with 98% accuracy 296 species of woody trees, shrubs, and palms of a forest plot [4]. The DNA barcode for phylogenetic applications in palms has recently been explored [5].

9.3 Markers for Palms from Other Monocot Plants [7–9]

About 96 nuclear "conserved intron-scanning primers" (CISP) were evaluated among the members of subfamilies of palms. Seven of them were useful in comparing subfamilies, and four of the loci have potential in comparing the related genera [6]. Nuclear CISPs were designed by the alignment of ESTs of onion and banana with BACs (bacterial artificial chromosome) of rice. Ninety-six of them gave a successful amplification rate of 32.3% in the DNA of members of the palm family [7]. Scarcelli et al. [8] developed about 100 new primers amplifying the selected regions in plastid

Genomics of Cultivated Palms. DOI: 10.1016/B978-0-12-387736-9.00009-1

loci of monocot genomes, which were used in some members of the palm family. A 70 kb sequence was obtained for Arecaceae plants by this process, which is equivalent to 44% of the known complete genome of the date palm plastid. In this sequence, 66 variable number of tandem repeats (VNTR) were found in palms, 77% of which were in the intergenic region, and the remaining in intronic sequences. The 51 polymorphic mononucleotide microsatellites encountered within palm genera and species can be compared with the 342 homopolymers of 7 bp or longer found in the complete chloroplast genome of *Phoenix dactylifera*.

9.4 Slow Evolution in Palms

A major problem in molecular phylogeny analysis of palms is the slow rate of evolution in commonly used DNA regions for plant phylogeny research. In comparing the *Adh* (alcohol dehydrogenase) gene sequences of palms and grasses [9], Morton et al. suggested the use of low-copy nuclear genes for phylogenetic research. Palm sequences [10] evolve at 2.61×10^{-9} substitutions per synonymous site per year, which is slower than most plant taxa. According to Gaut et al. [10], grass Adh genes evolve 2.5 times faster than palms at synonymous sites. Wilson, Gaut, and Clegg [11] analyzed the restriction variations of five enzymes (BamHI, ClaI, EcoR, HindIII, and HpaI) of 22 species of plastid DNA from five subfamilies of palms and found a 5- to 13-fold reduction in substitution rates within palm families relative to annual plant taxa. Further details of this topic are described by authors of [12].

9.5 Sequence Variations Used in Molecular Phylogeny of Plants [13–17]

Despite these limitations, a lot of work has been carried out using many nuclear and plastid DNA variations for phylogeny research in palms. DNA markers used in phylogeny of plants in general are RPB2 [13], ITS [14], and plastid regions trnT–F, rbcL [15], ndhF [16], and matK [15,17]. The RPB2 (RNA polymerase II) gene catalyzes the synthesis of mRNA in nuclei. It is a single-copy highly conserved gene and consists of 3.5 kb with 24 introns [13]. The length of the intron sequences displays polymorphism among species [13], making it the best candidate for phylogeny research.

Because of its rapid evolution, despite being a high copy number gene, the ITS (internal transcribed spacer) region [14] holds potential for plant phylogeny studies. Conserved regions of two spacer regions, ITS1 and ITS2, which have repeats, are useful in obtaining high-level relationships [14]. Of the plastid regions, the rapidly evolving matK (maturaseK) gene and the integeneric spacer region between the trnL and trnF (trnL–F) regions provide well-resolved and highly supported trees for other plants [15].

Table 9.1 Genomics of Phylogeny of Palms

Region	Type of Region, Location	Approximate Size (bp)	No. of PI Sites	Reference
rbcL	Coding, plastid	1500	92, 553, 55, 57	[18,19,21,22]
rps 16	Intron, plastid	680–960	74, 298, 47	[18,19,22]
trnL–F	Intergenic, plastid	740–950	63, 219, 57, 42	[18,19,20,22]
matK–trnK	Coding, intergenic, plastid	1800–1850	553	[19]
atpB	Coding, plastid	1300–1500	31	[22]

9.6 Plastid Regions in Classification of Palms

Three plastid regions (coding [rbcL] region, intron region [*rps* 16 intron], and intergeneric spacer region [trnL–F]) were analyzed for potential parsimony informative (PI) sites in palms [18]. In a study by Asmussen and Chase, the rbcL and *rps* 16 introns produced a high number of PI sites [18]. Comparative figures of PI sites in different plastid DNA regions [18–22], as reported by different researchers, are given in Table 9.1. The number of sites in different reports varies probably due to the number of species included in the study and sequence variation in the available set.

9.7 Nuclear DNA Regions in Phylogeny of Palms

Low-copy nuclear genes are good candidates for use in studying palm phylogeny. Specific coding regions of malate synthase (MS) exons were used for palms. The genes gave better phylogenetic resolution than noncoding plastid loci for palms at the family level. However, many clades were not resolved with sufficient bootstrap values [23]. RPB2 [13] and PRK are widely used in phylogenetic research of palms. Thomas et al. used the RPB and PRK sequence variation in concluding the monophyletic nature of the *Chamaedorea* genus [24] and the individuality of genera *Heterospathe* and *Rhopaloblaste* members of the Areceae taxon [25]. These two low-copy gene sequences also confirmed the many independent origins of the West Indian Palm flora [26]. The importance of RPB2 and PRK in classification of the Arecinae subtribe is dealt with in Chapter 6 under Section 6.2.

When used in 65 taxa, the nuclear rDNA 18 S SU gene sequence gave a topology similar to that of the plastid DNA-based study [22]. Sequences of WRKY transcription factors [27] were cloned in a set of palms of subtribe Attaleinae identified the genus *Syagrus* as a sister group of the coconut. Phytochrome genes, being repeated in plant families, are not suitable for phylogenetic research [28]. But recently, studies by Ludeña et al. [29] found AGA1 (Agamous 1) and phytochrome B (PHYB) to be single-copy genes in palms. The phylogenetic structure obtained with *AGA1* and *PHYB* genes clarified the intergeneric relationships within the Bactridinae. The

Hexopetion clade was found to be monophyletic with both markers [29]. The details are covered in Section 7.2. of Chapter 7.

Palm phylogeny was covered in detail earlier in Chapter 5, Section 5.2, on rattans Calamoideae group 5.2. Information on areca and related taxa is given in Chapter 6, Section 6.2, and on Bactridinae, in Chapter 7, Section 7.2. Chapter 8 deals with phylogeny research on the genera *Geonoma, Chamaedorea, Livistona, and Pinanga.*

An approach combining super tree and super matrix with DNA sequence-based data and morphological traits is now recognized to resolve palm phylogeny at the generic level [30]. Biogeography information is integrated with DNA sequence data to understand origin and diversification [31]. Such integrated approaches and modern tools (SNPs) will prove valuable in future palm phylogeny research.

References

1. Govaerts R, Dransfield J. World checklist of palms. Kew: Royal Botanic Gardens; 2005.
2. Dransfield J, Uhl NW, Asmussen NCB, Baker. WJ, Harley MM, Lewis CE. Genera palmarum: the evolution and classification of palms. Kew: Royal Botanic Gardens; 2008.
3. Anzizar I, Herrera M, Rohde W, Santos A, Dowe JL, Goikoetxea P, et al. Studies on the suitability of RAPD and ISTR for identification of palm species (Arecaceae). *Taxon.* 1998;**47**:635–645.
4. Kressa WJ, Ericksona DL, Jones FA, Swensond NG, Perezb R, Sanjurb O, et al. Plant DNA barcodes and a community phylogeny of a tropical forest dynamics plot in *Panama. Proc. Natl. Acad. Sci. USA*. 2009;**106**:18621–18626.
5. Jeanson ML, Labat JN, and Little DP (2011). DNA barcoding: a new tool for palm taxonomists? Ann. Bot. (in press). 10.1093/aob/mcr158.
6. Bacon CD, Feltus FA, Paterson AH, Bailey CD. Novel nuclear intron-spanning primers for Arecaceae evolutionary biology. *Mol. Ecol. Resour*. 2007;**8**:211–214.
7. Lohithaswa HC, Feltus FA, Singh HP, Bacon CD, Bailey CD, Paterson AH. Leveraging the rice genome sequence for monocot comparative and translational genomics. *Theor. Appl. Genet*. 2007;**115**:237–243.
8. Scarcelli N, Barnaud A, Eiserhardt W, Treier UA, Seveno M, d'Anfray A, et al. A set of 100 chloroplast DNA primer pairs to study population genetics and phylogeny in monocotyledons. *PLoS ONE*. 2011;**6**:e19954, 10.1371/journal.pone.0019954.
9. Morton BR, Brandon T, Gaut S, Clegg MT. Evolution of alcohol dehydrogenase genes in the palm and grass families. *Proc. Natl. Acad. Sci. USA*. 1996;**93**:11735–11739.
10. Gaut SB, Morton RB, Mccaig CB, Clegg TM. Substitution rate comparisons between grasses and palms: synonymous rate differences at the nuclear gene Adh parallel rate differences at the plastid gene rbcL. *Proc. Natl. Acad. Sci. USA*. 1996;**93**:10274–10279.
11. Wilson AM, Gaut SB, Clegg TM. Chloroplast DNA evolves slowly in the palm family (Arecaceae). *Mol. Biol. Evol.* 1990;**7**:303–314.
12. Baker WJ, Hedderson TA, Dransfield J. Molecular phylogenetics of subfamily Calamoideae (Palmae) based on nrDNA ITS and cpDNA rps16 intron sequence data. *Mol. Phylogenet. Evol.* 2000;**14**:195–217.
13. Denton AL, McConaughy BL, Hall BD. Usefulness of RNA Polymerase II coding sequences for estimation of green plant phylogeny. *Mol. Biol. Evol.* 1998;**15**:1082–1085.
14. Baldwin BG, Sanderson MJ, Porter JM, Wojciechowski MF, Campbell CS, Donoghue MJ. The ITS region of nuclear ribosomal DNA: a valuable source of evidence on angiosperm phylogeny. *Ann. Missouri Bot. Garden*. 1995;**82**:247–277.

15. Muller KF, Borsch T, Hilu KW. Phylogenetic utility of rapidly evolving DNA at high taxonomical levels: contrasting matK, trnT-F, and rbcL in basal angiosperms. *Mol. Phylogenet. Evol.* 2006;**41**:99–117.
16. Terry RG, Brown GK, Olmstead RG. Examination of subfamilial phylogeny in Bromeliaceae using comparative sequencing of the plastid locus ndhF. *Am. J. Bot.* 1997;**845**:664–670.
17. Kuo LY, Li FW, Chiou WL, Wnag CN. First insights into fern matK phylogeny. *Mol. Phylogenet. Evol.* 2011;**59**:556–566.
18. Asumussen CB, Chase MW. Coding and noncoding plastid DNA in palm systematic. *Am. J. Bot.* 2001;**98**:1103–1117.
19. Asmussen CB, Dransfiedl J, Deickmannm V, Barford AS, Pintaud JC, Baker WJ. A new subfamily classification of the palm family (Arecaceae): evidence from plastid DNA phylogeny. *Bot. J. Linn. Soc.* 2006;**151**:15–38.
20. Baker WJ, Asmussen CB, Barrow S, Dransfield J, Hedderson TA. A phylogenetic study of the palm family (Palmae) based on chloroplast DNA sequences from the trnL–trnF region. *Plant Syst. Evol.* 1999;**219**:111–126.
21. Hahn WJ. A phylogenetic analysis of the Arecoid line of palms based on plastid DNA sequence data. *Mol. Phylogenet. Evol.* 2002;**23**:189–204.
22. Hahn WJ. A molecular phylogenetic study of the palmae (Arecaceae) based on atpB, rbcL, and 18S nrDNA sequences. *Syst. Biol.* 2002;**51**:92–112.
23. Lewis CE, Doyle JJ. Phylogenetic utility of the nuclear gene malate synthase in the palm family (Arecaceae). *Mol. Phylogenet. Evol.* 2001;**19**:409–420, 10.1006/mpev.2001.0932.
24. Thomas MM, Garwood NC, Baker WJ, Henderson SA, Russell SJ, Hodel DR, et al. Molecular phylogeny of the palm genus Chamaedorea, based on the low-copy nuclear genes PRK and RPB2. *Mol. Phylogenet. Evol.* 2006;**38**:398–415.
25. Norup MV, Dransfield J, Chase MW, Barfod AS, Fernando ES, Baker WJ. Homoplasious character combinations and generic delimitation: case study from the Indo-Pacific arecoid palms. *Am. J. Bot.* 2006;**93**:1065–1080.
26. Roncal J, Zona S, Lewis CE. Molecular phylogenetic studies of Caribbean palms (Arecaceae) and their relationships to biogeography and conservation. *Bot. Rev.* 2008;**74**:78–102, 10.1007/s12229-008-9005-9.
27. Meerow AW, Noblick L, Borrone JW, Couvreur LP, Mauro-Herrera M, Hahn WJ, et al. Phylogenetic analysis of seven WRKY genes across the palm subtribe Attaleinae (Arecaceae) identifies Syagrus as sister group of the coconut. *PLoS ONE.* 2009;**4**(10):e:7353.
28. Mathews S, Lavin M, Sharrock RA. Evolution of the phytochrome gene family and its utility for phylogenetic analysis of angiosperms. *Ann. Mo. Bot. Garden.* 1995;**82**:296–321.
29. Ludeña B, Chabrillange N, Bertossi FA, Adam H, Tregear JW, Pintaud JC, et al. (2011). Phylogenetic utility of the nuclear genes AGAMOUS 1 and PHYTOCHROME B in palms (Arecaceae): an example within Bactridinae. *Ann Bot.* doi: 10.1093/aob/mcr191 (in press).
30. Baker WJ, Savolainen V, Asumussen-Lange CB, Chase MW, Dransfield J, Forest F, et al. Complete generic level phylogenetic analyses of palms (Arecaceae) with comparisons of supertree and supermatrix approaches. *Syst. Biol.* 2009;**58**:240–259.
31. Couvreur TLP, Forest F, Baker WJ. Origin and global diversification patterns of tropical rain forests: inferences from a complete genus level phylogeny of palms. *BMC Biol.* 2011;**9**:44.

About the Author

Dr. V. Arunachalam, born in India in 1969, received his B.Sc. (Horticulture) from Tamil Nadu Agricultural University, Coimbatore, India. He got his master's and doctoral degrees from Indian Agricultural Research Institute, Pusa, New Delhi, India.

He has 14 years of research experience in genetics and genomics of palms at the Central Plantation Crops Research Institute, Kasaragod, Kerala, India. Currently, he heads the horticulture section of ICAR Research Complex for Goa, Goa.

Dr. Arunachalam has guided many post-graduate students in their work on bioinformatics and biotechnology of palms, especially in coconut and arecanut research. He initiated a major externally funded program on computational genomics and the development of web tools for palms. He has a keen interest in bioinformatics and has worked on the discovery of new molecular markers from sequences. He has published many research articles and popular articles on palms. Learn more about him from the following websites: http://bioinfoindia.org/members/Arunachalam.htm and http://www.icargoa.res.in/arunachalam.htm.

CPSIA information can be obtained at www.ICGtesting.com
Printed in the USA
BVOW070437211211

278824BV00003B/93/P